HOW TO SPEAK
MACHINE

SIMPLICITY: DESIGN, TECHNOLOGY, BUSINESS, LIFE
John Maeda, Editor

The Laws of Simplicity, John Maeda, 2006

The Plenitude: Creativity, Innovation, and Making Stuff, Rich Gold, 2007

Simulation and Its Discontents, Sherry Turkle, 2009

Redesigning Leadership, John Maeda, 2011

I'll Have What She's Having, Alex Bentley, Mark Earls, and Michael J. O'Brien, 2011

The Storm of Creativity, Kyna Leski, 2015

The Acceleration of Cultural Change: From Ancestors to Algorithms, R. Alexander Bentley and Michael J. O'Brien, 2017

Mismatch, Kat Holmes, 2018

The Importance of Small Decisions: How Culture Evolves, Michael J. O'Brien, R. Alexander Bentley, and William A. Brock, 2019

Reimagining Design: Unlocking Strategic Creativity, Kevin G. Bethune, 2022

Nonlinear: Navigating Design with Curiosity and Conviction, Kevin G. Bethune, 2025

How to Speak Machine: A Gentle Introduction to Artificial Intelligence, with a new preface by the author, John Maeda, 2025

HOW TO SPEAK
MACHINE

A Gentle Introduction to Artificial Intelligence

with a new preface by the author

John Maeda

THE MIT PRESS
CAMBRIDGE, MASSACHUSETTS
LONDON, ENGLAND

The MIT Press
Massachusetts Institute of Technology
77 Massachusetts Avenue, Cambridge, MA 02139
mitpress.mit.edu

Reprinted by arrangement with Portfolio, an imprint of Penguin Publishing
Group, a division of Penguin Random House LLC.

The MIT Press would like to thank the anonymous peer reviewers who provided
comments on drafts of this book. The generous work of academic experts is
essential for establishing the authority and quality of our publications. We
acknowledge with gratitude the contributions of these otherwise uncredited
readers.

This book was set in Chronicle Text. Book design by NiCole LaRoche.
Printed and bound in the United States of America.

Library of Congress Cataloging-in-Publication Data

Names: Maeda, John, author.
Title: How to speak machine : a gentle introduction to artificial intelligence /
    John Maeda ; with a new preface by the author.
Description: Cambridge, Massachusetts : The MIT Press, [2025] | Series:
    Simplicity : design, technology, business, life | Includes bibliographical
    references and index.
Identifiers: LCCN 2024059768 | ISBN 9780262553179 (paperback)
Subjects: LCSH: Computer logic—Popular works. | Computer algorithms—
    Popular works. | Electronic data processing—Social aspects—Popular works. |
    Machine theory—Popular works. | Artificial intelligence—Popular works.
Classification: LCC QA76.9.L63 M34 2025 | DDC 006.3—dc23/eng/20241226
LC record available at https://lccn.loc.gov/2024059768

10   9   8   7   6   5   4   3   2   1

EU product safety and compliance information contact is: mitp-eu-gpsr@mit.edu

To my mother, Elinor "Yumi" Maeda, who taught me how to speak human. She taught me how to treat people the way they love to be treated—in the Hawaiian way.

To my father, Yoji Maeda, who taught me how to work like a machine. He taught me how to craft products that customers love and come back for—in the Japanese way.

Together they worked harder than any people I know so that I could attend any college in the world. Thanks to the two of them, I learned how to speak machine. And most importantly, thanks to Yumi, I haven't forgotten how to speak human too.

Contents

Preface

Dear Reader,

It's September 2024, and as I sit down to write this new preface, I'm struck by the eerie symmetry between the past and the present. In 2010, I wrote an essay for *Forbes* titled "Your Life in 2020," a speculative piece about the future of ubiquitous computing. It was a world where technology would fade into the background, becoming so seamless that it would be unnoticed—an era when "being digital" would transform into "having been digital." Now, as we stand on the cusp of 2025, I find myself reflecting on that vision, not just to see how accurate it was but to understand how we've arrived at this extraordinary moment in time—a moment dominated by artificial intelligence (AI) and large language models (LLMs).

Back then, the idea that technology would become invisible was both thrilling and unsettling. I envisioned a world where mobile phones would be as vital as the beating of our hearts, where STEM education—focused on science, technology, engineering, and math—would need an infusion of art to bring back a sense of humanity, evolving into what we now call STEAM. I imagined a software industry that would return to its craft heritage, moving away from mass-produced digital experiences to something more bespoke, more personal—something crafted by hand, even if that hand was a digital one.

Then the pandemic hit, and the world as we knew it changed overnight. What might have taken a decade to evolve in the digital space happened in mere months as businesses, schools, and entire industries were forced online. The pandemic became a catalyst for digital transformation—a shift driven by the unprecedented volumes of data generated during this time. By 2020, technology had indeed become pervasive, but rather than disappearing into the background, it became an omnipresent force, shaping our lives in ways that were both profound and, at times, disconcerting. What I hadn't fully anticipated was how quickly technology would evolve, and how it would bring us to the threshold of a new era—the era of AI.

As I look back at that essay from 2010, it's clear that we were on the brink of something massive, something that would change not just how we live and work but how we think about the very nature of human creativity. And yet, despite all the advances, one thing has remained constant: our desire to look at the past in order to better understand the future. At a 2005 commencement speech at Stanford, Steve Jobs famously said, "You can't connect the dots looking forward; you can only connect them looking backward. So you have to trust that the dots will somehow connect in your future." This reflection brings me to *How to Speak Machine*.

I spent six years writing *How to Speak Machine*, not just to demystify the world of computer science for the layperson but to reconnect with the foundational knowledge I had gained at MIT and in my career as a technologist. The process of writing the book was, in many ways, an attempt to reconcile the theoretical underpinnings of computer science with the realities of the rapidly evolving tech world I had been immersed in. It was a journey to understand how machines work but also how we, as humans, can better communicate with them—and, perhaps more importantly, how they can better communicate with us.

When *How to Speak Machine* was first published in 2019, the world was still grappling with the implications of AI. The idea that machines could "think" was both exhilarating and terrifying, depending on who you asked. At the time, large language models were still in their infancy, and the idea of designers using AI to automate their daily tasks seemed like something out of science fiction. But today, as we approach 2025, that science fiction has become science fact. Designers can indeed write code without engineers, using LLMs to generate and refine their work with a level of ease that would have been unimaginable just a few years ago.

This shift is particularly striking for me, having spent my early career teaching designers how to program, insisting they learn mathematics to fully grasp the logic behind the code. Today, they no longer need to wrestle with the complexities of programming languages because AI can write the code for them. Not only that, it can patiently explain it to them step by step, tailoring it to their respective levels of understanding. It's a golden age of craft, where the barriers between imagination and execution have dissolved, allowing the artistry of design to flourish in ways we only dreamed of back in 2010.

As we stand on the threshold of this new era, *How to Speak Machine* represents a pivotal moment—one that came just before the explosion of AI into the mainstream. This book served as a guide for me, helping me navigate the complexities of a world that was about to change in unimaginable ways. And now, as we move forward, it serves as a reminder of the importance of understanding our past to navigate our future. Just as Jobs's insight about connecting the dots by looking backward has guided me, I hope this book will guide you in understanding and shaping the technological landscape that lies ahead.

In the final chapter of *How to Speak Machine*, I concluded with a twist that might have surprised some readers. While the title of the book may have implied that it was about mastering the language of machines, the true invitation was something deeper: to prepare for a future where machines would become so dominant that we'd need to teach them how to speak human. The book wasn't just about speaking machine; it was about understanding the physics of computation so that when the time came, we could ensure that machines understood us—not just our language but our values, emotions, and humanity. This distinction is crucial. As we connect the dots from the past to the present, we see that the journey has been about more than just learning how machines work; it has been about preparing them to understand us on a human level.

The era of LLMs has only just begun, transforming industries in ways we're only beginning to comprehend. As I continue my work in AI now at Microsoft amid the disruptive force of ChatGPT, I'm reminded daily of the extraordinary capabilities of these systems—and the profound responsibilities they bring. We find ourselves in an era where machines seem to speak our language, but it's essential to remember that they don't yet speak human—they only process language as numerical tokens, devoid of true significance and accountability. To truly navigate this landscape, we must deepen our understanding of computational systems and how we've built tech products that appear to speak human but, in reality, speak only in patterns of data. We must not be deceived by this appearance. Only by mastering this distinction can we take on the greater mission of teaching machines to speak human in the fullest sense.

John Maeda
September 2024
Redmond, Washington

Introduction

It was a typical cold, snowy New England winter day on December 17, 2004, when I started a WordPress "web log" on the topic of simplicity. There was a whimsical motivation to turning my research at MIT in that direction—namely, how the letters MIT occurred in perfect sequence in the words SIMPLICITY and COMPLEXITY. But there was a less Dr. Seussian motivation at play as well, which I wrote about in my first blog entry:

> I have always been interested in how the computer (which is an object of great complexity) and design (which is traditionally about simplicity) tend to mix poorly together like the proverbial "oil and water."

Subsequently, that blog turned into a book titled *The Laws of Simplicity*, which was rapidly translated into fourteen languages. Why was it unusually impactful? I think because it arrived at a time when computing technology was just starting to impact everyday lives back in the pre-iPhone era. That book's overwhelming momentum and the concurrent rise of Apple's successful fusing of design and technology oddly drove me to head in the opposite direction of computing's inherent complexities and instead toward designing for simplicity.

I wanted to somehow get closer to the essence of design and move away from computers the way I had done once prior in my early career—back in the nineties, when I was a practicing graphic designer in Japan with a mismatched MIT pedigree. I'd somehow managed to escape the "T" (Technology) of MIT as an engineering student, and then made a U-turn into the thick of it as an MIT Media Lab professor leading the intersection of design and advanced computing technologies. Perhaps it was dealing with the weight of earning tenure that made me feel stuck somewhere in the future of design. I wanted to reconnect with the classics. I think it was a mix of not knowing what to do with the MBA I'd earned as a part-time hobby and an overwhelming mood in 2008 of Barack Obama's "Yes, we can," combined with my desire to rekindle the past, that resulted in my becoming the sixteenth president of the esteemed "temple" of the art and design world: Rhode Island School of Design, or RISD.

A series of later milestones, which built upon the work I had long led at the MIT Media Lab, gave me a reputation as a fierce defender of design. These included, for instance, appearing before Congress to encourage putting an "A," for Art, in STEM education, turning it into STEAM, and later launching the "Design in Tech Reports" while working in Silicon Valley at venture capital firm Kleiner Perkins. So in 2019, when a popular business magazine announced in a headline that I'd said, "In reality, design is not that important," it did not come as a surprise to me that I would be dragged through the internet mud by all lovers of design.

There were many short- and long-form responses to the interview that cast my responses as every shade of clueless and ignorant. I knew that the controversy had hit a peak when my idol,

Hartmut Esslinger, the design force behind Apple's original design language, started coming after me on social media. If I had earned some sort of "badge" in the design world, the internet was now ruling that it was my public duty to turn that badge in, and for a period of time I would be unwelcome at any Temple of Design out there. How did I feel? Terrible.

My words had been taken out of context from a twenty-minute phone interview—and, frankly, when the article came out, I immediately admired the editorial team's choice of headline as brilliant clickbait. Apparently, my interview was the highest-performing article on their website for quite some time, as evidenced by the myriad spears and cleavers that were continuously being lobbed in my direction. What stung the most was knowing that few people had really read the entire interview, so the headline was all that stuck in anyone's mind. To them, I had completely demeaned the work designers do every day. So I needed to be punished.

Here's the reality: I honestly don't believe that design is the most important matter today. Instead, I believe we should focus first on understanding computation. Because when we combine design with computation, a kind of magic results; when we combine business with computation, great financial opportunities can emerge. What is computation? That's the question I would get asked anytime I stepped off the MIT campus when I was in my twenties and thirties, and then whenever I left any technology company I worked with in my forties and fifties.

Computation is an invisible, alien universe that is infinitely large and infinitesimally detailed. It's a kind of raw material that doesn't obey the laws of physics, and it's what powers the internet at a level that far transcends the power of electricity. It's a ubiquitous medium that experienced software developers and the tech industry control to a degree that threatens the sovereignty of existing nation-states. Computation is not something you can fully grasp after training in a "learn to code" boot camp, where the mechanics of programming can be easily learned. It's more like a foreign country with its own culture, its own set of problems, and its own language—but where knowing the language is not enough, let alone if you have only a minimal understanding of it.

There's been a conscious push in all countries to promote a greater understanding of how computers and the internet work. However, by the time a technology-centered educational program is launched, it is already outdated. That's because the pace of progress in computing hasn't moved at the speed of humans—it's been moving at the exponential speed of the internet's evolution. Back in 1999, when a BBC interviewer made a dismissive comment about the internet, the late musician David Bowie presciently offered an alternate interpretation: "It's an alien life form . . . and it's just landed here." Since the landing of this alien life form, the world has not been the same—and design as it has conventionally been defined by the Temple of Design no longer feels to me like the foundational language of the products and services worlds. Instead, it's ruled by new laws that are governed by the rising Temple of Tech in a way that intrinsically excludes folks who are less technically literate.

A new form of design has emerged: computational design. This kind of design has less to do with the paper, cotton, ink, or steel that we use in everything we physically craft in the real world, and instead has more to do with the bytes, pixels, voice, and AI that we use in everything we virtually craft in the digital world powered by new computing technologies. It's the text bubble that pops up on your screen with a message from your loved one, or the perfect photo you shot in the cold rain with your hands trembling and yet which came out perfectly, or the friendly "Here you go, John" that you hear when you ask your smart stereo to play your favorite Bowie tunes. These new kinds of interactions with our increasingly intelligent devices and surroundings require a fundamental understanding of how computing works to maximize what we can make.

So I came to wonder if I could find a way for more nontechie people to start building a basic understanding of computation. And then, with that basic conceptual grounding, to show how computation is transforming the design of products and services. For much of the twentieth century, computation by itself was useful only to the military to calculate missile trajectories. But in the twenty-first century, it is *design* that has made computation relevant to business and, more so, to our everyday lives. Design matters a lot when it is leveraged with a deep understanding of computation and the unique set of possibilities it brings. But achieving an intuitive understanding of an invisible alien universe doesn't come so easily.

This book is the result of a six-year journey I have traveled away from "pure" design and into the heart of what is impacting design the most: computation. I will take you on a tour through the minds and cultures of computing machines from how they once existed in a simpler form to how they've evolved into the much more complex forms we know today. Keep in mind that this book is not designed to turn you into a computer science genius—I've vastly simplified, and in some cases oversimplified, technical concepts in a way that will surely raise some experts' eyebrows and might cause them to cringe. But I hope that, armed with even my rough approximations, you will learn how computation has expanded both in technical capability and in sociocultural impact—something you may regard simultaneously as hugely impressive and terribly frightening.

Computation brings its share of problems, but most of them have to do with us—how we use it—rather than by the underlying technology itself. We've entered an era in which the computing machines we use today are powered not just by electricity and mathematics, but by our every action and with insights gained in real time as we use them. In the future we'll have only ourselves to blame for how computation evolves, but we're more likely to succumb to a victim mentality if we remain ignorant as to what is really going on. So it will become only natural to want to pin the blame on a handful of tech company leaders, if not all of them—a fairly likely scenario, since fear of the invisible or unknown is far more powerful than any fear that has physical form, like a pack of wild wolves or a threatening tornado. The intangible, invisible alien force that is the internet represents the perfect object of fear that already lives in your neighborhood and teaches your children while secretly seeking to do you all harm, which explains why the eeriness of tech is so widely portrayed in TV and movies today.

I have always believed that being curious is better than being afraid—for when we are curious we get inventive, whereas when we are afraid we get destructive. Something about my experience between the Temple of Design and the Temple of Tech has kept me curious all these years. When I think more deeply about it, it's the many career failures I've fortunately experienced in between my few successes that drive me to still remain a little hungry and foolish. But to be honest, I'm just like anyone else—tired, a little lazy, and all too eager to wait for a hero to rise who will protect me and fight for all of us. There's a common lack of understanding of what computation fundamentally can and cannot do. Rather than give away your power of understanding to someone else, I invite you to be curious about the computational universe.

Perhaps I wrote this book for you. Perhaps *you* are the hero the world has been waiting for. Perhaps you are one of the many who will find a way to wield the power of computation with inventiveness and wonder. Those kinds of heroes are now desperately needed in order to advance computation beyond what it is today in its superpowerful, albeit running with the conflicted conscience of a teenager, form. Being new to the computational universe, you just might discover something that we first-generation techies have not yet been able to imagine. When you find it and make it into a success, it will set an example for the rest of us. I wish that heroic moment upon you someday, but first let me start you on the path to speaking the language of the machine.

Chapter 1

MACHINES RUN
LOOPS

1 // Computers excel at repeating themselves through loops.

2 // Hard machines are visible; soft machines are invisible.

3 // Human computers are the original computing machines.

4 // Recursion is the most elegant means to repeat oneself.

5 // Loops are indestructible unless a programmer has made an error.

1. Computers excel at repeating themselves through loops.

Physical education was never my forte growing up. Not only was I the second-fattest kid in class, but I also couldn't for the life of me throw or catch a ball of any size. My one special power was staying awake for a long time, thanks to the influence of my father and the boot camp mentality with which he ran the family business. But that little bit of physical pride meant nothing when running loops around the school and constantly getting lapped by all my classmates. I always felt it was so boring, not to mention tiring, to run those loops around the school grounds. Although I wasn't particularly good at it, I knew I wasn't the only exhausted one—even the fastest runners looked winded by the time I got to the finish line. Athletic or not, we're all animals that, no matter how physically fit, will tire eventually.

There's one thing that a computer can do better than any human, animal, or machine in the real world: repetition. It doesn't get bored and complain if you tell it to count from one to a thousand, or even to a billion. You just tell it to start at zero, add one, and repeat until your target number is reached. And the computer just goes off and does it. For example, I type in these three lines of code into my computer:

```
top = 1000000000
i = 0
while i < top: i = i + 1
```

which compels the computer to count to one billion, within one minute, and it's then waiting for me to give it a new set of instructions to execute under my complete control. It's eager to please me.

Let's think about that for a second. The hamster on the treadmill spins it round and round fast, but eventually tires. The Formula One racing car races the track at high speeds but eventually runs out of gas and needs to stop. If the hamster were to not stop, we would be rather concerned. Our first thought would be, *Freaky!* If the F1 race continued with no need for the cars to make a pit stop, we'd think to ourselves, *Magic!*

But a computer running a program, if left powered up, can sit in a loop and run forever, never losing energy or enthusiasm. It's a metamechanical machine that never experiences surface friction and is never subject to the forces of gravity like a real mechanical machine—so it runs in complete perfection. This property is the first thing that sets computational machines apart from the living, tiring, creaky, squeaky world.

I first came in touch with the power of computational loops in 1979, when I was in seventh grade. I had just encountered my first computer. That in itself was unusual for someone with my background growing up on the poor side of town; thanks to desegregation efforts motivated by the civil rights movement, I was bused to a school an hour away from my home that was much better than the run-down schools in my neighborhood.

Commodore was a big name in computing at the time. When I say big, of course I mean big for maybe a few thousand people in the United States and Europe. Personal computers weren't really personal back then, because the average family could not afford one.

The Commodore PET (Personal Electronic Transactor) was manufactured in the United States with a small screen displaying text in only fluorescent green, a tiny tactile keyboard, and a tape cassette drive for storage. It had 8 kilobytes of total memory, and its processing speed was 1 megahertz. In comparison, the average mobile phone has 8 gigabytes and runs at 2 gigahertz, which is a millionfold increase in memory and a thousandfold increase in speed.

There was no internet yet, so you couldn't search for anything. There was no Microsoft, so you couldn't do your schoolwork on a word processor or spreadsheet. There was no touch screen or mouse, so you couldn't directly interact with what was on the monitor. There were no color or grayscale pixels to display images with, so you couldn't visually express information. There was only one font, and text was only uppercase. You would "navigate" the computer screen by pressing the cursor keys: up, down, left, right. And any functionality that you wanted it to have, you would need to type in a program to create it yourself or type it in line by line from a book or magazine.

As you can imagine, the computer sat in the classroom generally unused—it was not only useless, it was soulless as an experience. No expressive or informative images. No stereo sound or the latest tunes. No utility or empowerment with an amazing set of apps. It just blinked at you, constantly, with its cursor rectangle—awaiting you to type in instructions for it to follow. And when you did raise the courage to type something into it, you would likely be rewarded with, in all caps, SYNTAX ERROR—which essentially translated to YOU'RE WRONG. I DON'T UNDERSTAND.

It's no surprise that the computer attracted only a few kinds of students—perhaps those who grew up a bit lower on the empathy scale (like me), or those who could tolerate the crushing blow of being told they were wrong with each keystroke. My friend Colin, whose parents worked with computers at Boeing, showed me my first program. He rapidly typed the following program into the PET, free from any syntax errors:

```
10 PRINT "COLIN"
20 GOTO 10
```

Then he told me to go ahead and type RUN . . . What happened next astounded me. The computer began printing "COLIN" continuously. I asked when it was going to stop. Colin said, "Never." This worried me. He then proceeded to break the program with CONTROL-C. And then the blinking text prompt came back.

Colin then retyped the first line of code, but this time with one space and a semicolon.

```
10 PRINT "COLIN ";
```

And he typed RUN to display the following:

```
COLIN COLIN COLIN COLIN COLIN COLIN COLIN COLIN COLIN
COLIN COLIN COLIN COLIN COLIN COLIN COLIN COLIN COLIN
COLIN COLIN COLIN COLIN COLIN COLIN COLIN COLIN COLIN
COLIN COLIN COLIN COLIN COLIN . . .
```

And again, it kept printing and scrolling by. I tried it myself, and typed:

```
10 PRINT "MAEDA " . . .
```

to experience the incredibly affirming display of my name being said forever and ever:

```
MAEDA MAEDA MAEDA MAEDA MAEDA MAEDA MAEDA MAEDA MAEDA
MAEDA MAEDA MAEDA MAEDA MAEDA MAEDA MAEDA . . .
```

Thereafter, I would perform this magic "say my name" trick for anyone who was curious about the computer. I did it for my classmate Jessica, who I kind of had a thing for. The limits of my computer expertise became evident when she asked, "What else can you do?" *Uh-oh.*

But my curiosity was piqued. I started to read *Byte* magazine (one of maybe two computing magazines available). Since there was close to no software available, it was really important to learn to write programs. *Byte* regularly included entire computer programs printed out across many pages and ready to be manually typed in to a computer yourself—the only problem was that I didn't have a computer to regularly use.

Luckily, my mother, Elinor, was always forward-looking and hopeful that her children could do bigger and better things in life. She set aside enough money from our small, family-operated tofu shop in Seattle to buy me an Apple II computer and an Epson line printer. To express my gratitude to her, I wanted my first computer program to help her in some way at the tofu shop. Thus I set out to write a monthly billing program that I hoped could save her some time. It would manually take in our regular customers' orders each week and print out an invoice at the end of the month.

I was a fast typist by the tenth grade, and so with zeal I wrote this program that I felt could help my mother. It took me maybe three months of programming every day after school. My conundrum was figuring out how to deal with leap years—I figured that if I made an input routine for 365 days in a year, I would run into a problem every four years. In the end I saw that as a bridge to cross when I came to it, so instead I just kept typing and typing until I had completed all 365 input routines (there were no text editors with copy and paste functions). It was a manual, laborious project. I recall the deep satisfaction I felt when my mother first used it to print invoices for the month.

Shortly after this moment of success, my tenth-grade math teacher, Mr. Moyer, encouraged me to come to his after-school computer club. I had gained a reputation for being skilled at computer programming—perhaps what I should really say is that I was a computer nerd. Having successfully written my first thousand-line computer program, I thought it would be beneath me to go to Mr. Moyer's club meeting as I would surely be too much of an expert compared with the others attending. But on the day I showed up, I vividly remember Mr. Moyer talking about LOOPS using a command called FOR . . . NEXT. Upon learning about it, I broke into a sweat—the kind of sweat you feel when you've done something terribly stupid.

When I got home that evening, I looked at my long computer program, which was 365 distinct input statements of the form:

```
10 DIM T(365), A(365) : HOME
100 REM GET THE NUMBER OF TOFU AND SUSHI AGE FOR EACH
    DAY OF THE YEAR
110 REM COMMENTS LIKE THIS ARE HOW PROGRAMMERS TALK TO
    THEMSELVES
```

```
120 PRINT "IT'S DAY 1"
130 PRINT "HOW MANY TOFU"
140 INPUT T(1)
150 PRINT "TOFU ORDER IS", T(1)
160 PRINT "HOW MANY DOZEN SUSHI AGE"
170 INPUT A(1)
180 PRINT "SUSHI AGE ORDER IS", A(1)
290 PRINT "CONTINUE? HIT 0 TO EXIT OR 1 TO CONTINUE"
200 INPUT ANSWER
210 IF (ANSWER = 0) GOTO 9999
220 PRINT "IT'S DAY 2"
230 PRINT "HOW MANY TOFU"
240 INPUT T(2)
250 PRINT "TOFU ORDER IS", T(2)
260 PRINT "HOW MANY DOZEN SUSHI AGE"
270 INPUT A(2)
280 PRINT "SUSHI AGE ORDER IS", A(2)
290 PRINT "CONTINUE? HIT 0 TO EXIT OR 1 TO CONTINUE"
300 INPUT ANSWER
310 IF (ANSWER = 0) GOTO 9999
320 PRINT "IT'S DAY 3"
330 PRINT "HOW MANY TOFU"
340 INPUT T(3)
350 PRINT "TOFU ORDER IS", T(3)
360 PRINT "HOW MANY DOZEN SUSHI AGE"
370 INPUT A(3)
380 PRINT "SUSHI AGE ORDER IS", A(3)
390 PRINT "CONTINUE? HIT 0 TO EXIT OR 1 TO CONTINUE"
400 INPUT ANSWER
```

```
410 IF (ANSWER = 0) GOTO 9999
420 PRINT "IT'S DAY 4"
430 PRINT "HOW MANY TOFU"
440 INPUT T(4)
450 PRINT "TOFU ORDER IS", T(4)
460 PRINT "HOW MANY DOZEN SUSHI AGE"
470 INPUT A(4)
480 PRINT "SUSHI AGE ORDER IS", A(4)
490 PRINT "CONTINUE? HIT 0 TO EXIT OR 1 TO CONTINUE"
500 INPUT ANSWER
510 IF (ANSWER = 0) GOTO 9999
520 PRINT "IT'S DAY 5"
530 PRINT "HOW MANY TOFU"
540 INPUT T(5)
550 PRINT "TOFU ORDER IS", T(5)
560 PRINT "HOW MANY DOZEN SUSHI AGE"
570 INPUT A(5)
580 PRINT "SUSHI AGE ORDER IS", A(5)
590 PRINT "CONTINUE? HIT 0 TO EXIT OR 1 TO CONTINUE"
600 INPUT ANSWER
610 IF (ANSWER = 0) GOTO 9999
620 REM CONTINUE SIMILARLY FOR 360 MORE TIMES BY TYPING
    AS FAST AS YOU CAN
9999 PRINT "ALL DATA ENTERED"
```

and so on and so on for 360 more times . . .

I was referring to days of the year as numbers from 1 to 365 and had strung them together as a long program. We sold five or six products in addition to tofu and sushi age (*abura-age*), and I included a lot more dialogue text. So it was plenty to type for each day of the year. And there were many GOTO statements to route the logic in ways that could give greater clarity to my mother as she input the data that I'd left out in the above depiction.

I then rewrote the program with my newly learned technique from Mr. Moyer:

```
10 DIM T(365), A(365) : HOME
100 REM THANK YOU, MISTER MOYER
110 FOR I = 1 TO 365
120 PRINT "IT'S DAY", I
130 PRINT "HOW MANY TOFU"
140 INPUT T(I)
150 PRINT "TOFU ORDER IS", T(I)
160 PRINT "HOW MANY DOZEN SUSHI AGE"
170 INPUT A(I)
180 PRINT "SUSHI AGE ORDER IS", A(I)
190 PRINT "CONTINUE? HIT 0 TO EXIT OR 1 TO CONTINUE"
200 INPUT ANSWER
210 IF (ANSWER = 0) GOTO 9999
220 NEXT
230 REM NO NEED TO TYPE ANY MORE ENTRIES. RELAX!
9999 PRINT "ALL DATA ENTERED"
```

Done!

I stared incredulously at the 365 separate programming chunks, which comprised 40 or so lines of programming per entry point—totaling roughly 14,600 lines of code. In under half an hour I was down to less than 50 lines of code. My ego was appropriately deflated. Up until that moment, I took pride in my ability to get work done by sheer brute force—manually. But I realized that if I instead could think in LOOPS the way a computer natively thinks, it could get my work done with elegance—automatically. It just required me to learn the right way to formulate a repetitive task for the computer. Then I could just wind it up like a toy mechanical car, and off it goes on its own!

Making the computer do the same thing over and over seems like we're taking advantage of its lack of intelligence. But we must exercise a great deal of our own intelligence to turn repetition into a form of art in code. Make no mistake: the machine speaks a foreign language, with its own vocabulary and grammar, and it will take you more than reading this short book to speak machine fluently. I can, however, help you get the gist of computation, and to do so I'll take an important detour into the essence of what software is all about.

2. Hard machines are visible; soft machines are invisible.

In the early 2000s, when I was at the MIT Media Lab, I was hanging out with an executive from Motorola discussing the iconic StarTAC flip phone and its amazing success in popularizing mobile telephony. I was nothing but enthusiastic about the phone's potential. The executive was not. Sales for mobile handsets were starting to stall at Motorola, a fact he explained with a prediction that came true and has always stuck with me. In the old days, he said, you could count on shipping an amazing hardware device that would come packaged with a CD-ROM that you'd easily toss into the trash. But the day was soon coming when you'd be doing the opposite: coveting the software and tossing out the hardware instead. In essence, he was describing the era we live in today, in which there are apps for everything and we depend more on the software than the actual physical devices in which they function.

On the surface, most software has a visual representation that we mistakenly identify as the actual computational machine. What you see on-screen with an app is closer to the fast-food drive-thru sign that you drive up to and speak into—which has nothing inside it except a tethered connection to a bustling food factory just a few car lengths away. Just as you can learn nothing about how the actual restaurant works by taking apart the drive-thru's microphone box, the pixels on a computer screen don't tell us anything about the computational machine that it is connected to. Contrast that with cracking open the hardware you're running that app on—although its internals would be a bit confusing, you would still find the screen, the battery or power supply, and a few other recognizable parts. That's because when it comes to something that exists in the physical world, you can touch and understand it, to a degree, when you crack it open. Machines in the real world are made up of wires, gears, and hoses that kind of make sense, whereas machines in the digital world are made up of "bits" or "zeroes and ones," which are completely invisible.

But what about the program codes from the previous section,

where I produced an invoice for my parents' tofu shop? The software, written in BASIC, is something you can read with your eyes or ears. Is software visible? Yes and no. On the one hand, program code is what lies at the heart of software and you can read it, but that's like confusing the recipe for cake with the cake itself. The software is what comes alive inside the machine due to the program codes—it's the cake, not the recipe. This can be a difficult conceptual leap.

Understanding the distinction between software running on a computer within its "mind" versus the actual program code being fed into the computer is useful because it lets you conceptualize what is really happening with computation. It can free you from believing that computer code is just, well, computer code—which on the surface is all you can generally see. But what computer code can represent is where the true potential lies. We could say the same for the words you are reading on this page in the sense that they spark intangible ideas in your mind, so it's not the actual words you are experiencing but the invisible ideas that underlie them instead. And in the same way that you know how powerful your imagination can become when fed with the right literary fuel (hopefully that includes this book), then your mind is empowered to do things you previously thought were impossible. That's what happens when a well-crafted computer program is brought to life with a finger tap or a double click—an alternate, invisible consciousness instantly manifests, just like the magical moment when hydrating a completely desiccated sponge.

Computing machines can freely imagine within "cyberspace," a term coined by William Gibson in the novel *Neuromancer* in 1984, the same year I started at MIT:

Cyberspace, a consensual hallucination, experienced daily by billions of legitimate operators, in every nation, by children being taught mathematical concepts . . . Unthinkable complexity. Lines of light ranged in the nonspace of the mind, clusters and constellations of data. Like city lights receding . . .

Trippy. But accurate, or at least close to what I've viscerally experienced within the invisible world "inside" the machine when writing code—note that the quotes around the word "inside" are important because there's nothing that lies within the visible shell of an app. There's a nether universe that computational machines can easily tap into that precedes the internet, and now because of the internet and the ubiquity of network-enabled devices, that universe has expanded wildly beyond what any of the lucky nerds like me who were there at the beginning could ever have expected. Gibson's "consensual hallucination, experienced daily by billions," can on the one hand allude to Facebook or any social media network today, or a shared multiplayer video game in a lush three-dimensional virtual world—or else in the less concrete direction of the "unthinkable complexity" that Gibson refers to, which is a better characterization of what I conjure within the poetry of his description of "lines of light ranged in the nonspace of the mind, clusters and constellations of data."

As you can tell, I'm unusually passionate about this subject, and quite eager for you to understand it with me. Whether by creating an art installation in Kyoto in 1993 to try to visualize the inner workings of a computer as a literal discotheque of people posing as computer parts, or by projecting on nine large screens in a dark gallery billions of chaotic particles buzzing about like bees for my 2005 exhibition at the Cartier Foundation in Paris, I've wanted more people to experience what digital consciousness can feel like. Why? Because I believe to speak machine, you need to "live" the world of the machine too. And, unfortunately, it is by nature invisible.

One way to understand it is by becoming a master computer programmer, but that's not the desired path for everyone. So as we move forward through the chapters of the book, try to keep Gibson's image of cyberspace as an apt representation of how native machine speakers collectively "see" the invisible.

Before we make the jump fully back into cyberspace, let's briefly dip into the topic of the final chapter of this book—Machines Automate Imbalance—to examine another invisible aspect of the history of computation that will be of great benefit to know. Just like any efficient computational machine, any given history of human beings will be repeated over and over until it becomes perceived as fact. So before you get too excited about machines, let's embrace more of the invisible by looking back at when human beings were the fully visible machinery of computation. And in the process you will have the opportunity to rewrite computing's history by properly including the many professional women who were unfairly made invisible.

3. Human computers are the original computing machines.

The first computers were not machines, but humans who worked with numbers—a definition that goes back to 1613, when English author Richard Braithwaite described "the best arithmetician that ever breathed" as "the truest computer of times." A few centuries later, the 1895 *Century Dictionary* defined "computer" as follows:

> One who computes; a reckoner; a calculator; specifically, one whose occupation is to make arithmetical calculations for mathematicians, astronomers, geodesists, etc. Also spelled computor.

At the beginning and well into the middle of the twentieth century, the word "computer" referred to a person who worked with pencil and paper. There might not have been many such human computers if the Great Depression hadn't hit the United States. As a means to create work and stimulate the economy, the Works Progress Administration started the Mathematical Tables Project, led by mathematician Dr. Gertrude Blanch, whose objective was to employ hundreds of unskilled Americans to hand-tabulate a variety of mathematical functions over a ten-year period. These calculations were for the kinds of numbers you'd easily access today on a scientific calculator, like the natural exponent e^x or the trigonometric sine value for an angle, but they were instead arranged in twenty-eight massive books used to look up the calculations as expressed in precomputed, tabular form. I excitedly purchased one of these rare volumes at an auction recently, only to find that Dr. Blanch was not listed as one of the coauthors—so if conventional computation has the problem of being invisible, I realized that human computation had its share of invisibility problems too.

Try to imagine many rooms filled with hundreds of people with a penchant for doing math, all performing calculations with pencil and paper. You can imagine how bored these people must have been from time to time, and also how they would have needed breaks to eat or use the bathroom or just go home for the evening. Remember, too, that humans make mistakes sometimes—so someone who showed up to work late after partying too much the night prior might have made a miscalculation or two that day. Put most bluntly, in comparison with the computers we use today, the human computers were comparatively slow, at times inconsistent, and would make occasional mistakes that the digital computer of today would never make. But until computing machines came along to replace the human computers, the world needed to make do. That's where Dr. Alan Turing and the Turing machine came in.

The idea for the Turing machine arose from Dr. Turing's seminal 1936 paper "On Computable Numbers, with an Application to the Entscheidungsproblem," which describes a way to use the basic two acts of writing and reading numbers on a long tape of paper, along with the ability to write or read from anywhere along that tape of paper, as a means to describe a working "computing machine." The machine would be fed a state of conditions that would determine where the numbers on the tape would be written or rewritten based on what it could read—and in doing so, calculations could be performed. Although an actual computing machine could not be built with technology available back then, Turing had invented the ideas that underlie all modern computers. He claimed that such a machine could universally enable any calculation to be performed by storing the programming codes onto the processing tape itself. This is exactly how all computers work today: the memory that a computer uses to make calculations happen is also used to store the computer codes.

Instead of many human computers working with numbers on paper, Alan Turing envisioned a machine that could tirelessly calculate with numbers on an infinitely long strip of paper, bringing the exact same enthusiasm to doing a calculation once, or 365 times, or even a billion times—without any hesitation, rest, or complaint. How could a human computer compete with such a machine? Ten years later, the ENIAC (Electronic Numerical Integrator and Computer), built for the US Army, would be one of the first working computing machines to implement Turing's ideas. The prevailing wisdom of the day was that the important work of the ENIAC was the creation of the hardware—that credit being owned by ENIAC inventors John Mauchly and John Presper Eckert. The perceived "lesser" act of programming the computer—performed by a primary team of human computers comprising Frances Elizabeth Snyder Holberton, Frances Bilas Spence, Ruth Lichterman Teitelbaum, Jean Jennings Bartik, Kathleen McNulty Mauchly Antonelli, and Marlyn Wescoff Meltzer—turned out to be essential and vital to the project, and yet the women computers of ENIAC were long uncredited.

As computation could be performed on subsequently more powerful computing machines than the ENIAC and human computers started to disappear, the actual act of computing gave way to writing the set of instructions for making calculations onto perforated paper cards that the machines could easily read. In the late 1950s, Dr. Grace Hopper invented the first "human readable" computer language, which made it easier for people to speak machine. The craft of writing these programmed instructions was first referred to as "software engineering" by NASA scientist Margaret Hamilton at MIT in the 1960s. Around this time, Gordon Moore, a pioneering engineer in the emerging semiconductor industry, predicted that computing power would double approximately every year, and the so-called Moore's law was born. And a short two decades later I would be the lucky recipient of a degree at MIT in the field that Hamilton had named, but with computers having become by then many thousands of times more powerful—Moore's exponential prediction turned out to be right.

To remain connected to the humanity that can easily be rendered invisible when typing away, expressionless, in front of a metallic box, I try to keep in mind the many people who first served the role of computing "machinery," going back to Dr. Blanch's era. It reminds us of the intrinsically human past we share with the machines of today. In the book of tabulations by Dr. Blanch's team that I own—*Table of Circular and Hyperbolic Tangents and Cotangents for Radian Arguments,* which spans more than four hundred pages, with two hundred numbers calculated to seven decimal places—I find it humanly likely that a few of the calculations printed on those pages are incorrect due to human error. It was humans who built the Turing machines that have eradicated certain kinds of human error and who have made it possible to speak with machines in many fanciful computer languages. But it is all too easy to forget that humans err all the time—both when we are the machines and when we have made the machines err on our behalf. Computation has a shared ancestor: us. And although historically most of our mistakes made with computers have been errors in straightforward calculations, we now need to come to terms with the mistaken human assumptions that are embedded in our calculations, like the countless omissions in history of the role of women in computing. Computation is made by us, and we are now collectively responsible for its outcomes.

4. Recursion is the most elegant means to repeat oneself.

When I arrived at MIT as an undergraduate in 1984, computer science was just starting to take off as a field, championed by a new textbook written by Hal Abelson, Gerald Sussman, and Julie Sussman. They were lifting the idea of computer programming out of its craft heritage and advancing its practice into the world of critical thinking and science. They were not alone in this endeavor—it was a movement occurring across the world, but primarily rooted in universities because, frankly, nobody else had (or wanted) computers at the time. There was no Google, Yahoo, Amazon, Facebook, etc. And don't forget that it was back at a time when people would laugh at you (me) for having a weird-looking contraption with a rainbow-colored fruit on it.

Abelson and Sussman's introductory course to computer

science was titled Structure and Interpretation of Computer Programs. We were the first year of students to use their newly printed book of the same name, and it was way, way, *way* over my head every week. Their goal was to get us to think differently. In week one of their course, we were told to no longer think of the world in terms of GOTO 10 or FOR . . . NEXT loops. They introduced us to a new concept called "recursion"—which is one of those ideas that you don't naturally come across in your daily life, because it doesn't fit how the physical world around us usually works. It's not just a little strange—it's extremely strange. Let me try to explain this more concretely.

As a child, I was terribly superstitious and believed in magic. I guess today I still do. It all started when I discovered that I had the power to see through my hand like Superman. You can do this yourself by having both eyes open and placing one hand in front of one of your eyes about a foot away. If you're like me, you'll be able to see through that hand and everything that's behind it. It's something I never wanted to tell anybody about because it felt a bit creepy to have this special power.

So when I found, in the teen romance comic *Archie* (I know, I was a bit precocious for my young age), an introduction to a Möbius strip explained on its crafts page, and when I actually made one, I felt my magical powers truly starting to manifest. The Möbius strip is a simple object made out of folded paper that reveals surprising properties as you start to play with it.

Take two strips of paper and make two loops: one just by connecting it end to end, and the other by twisting the end connection just once before you seal the loop. The first strip is just a regular loop. The second is a Möbius strip.

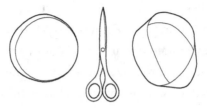

Now grab a pair of scissors and cut the regular loop down the middle. What happens? You get . . . two loops. That makes sense, because you've chopped the loop in half. But when you cut the Möbius strip down the middle, what happens? You don't get two loops—you get one loop instead.

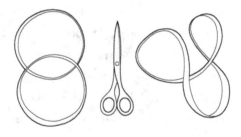

You can imagine that, as a kid believing in magic, I totally flipped out. It felt like I'd stumbled upon some sort of dark magic. I wondered if I had unleashed the same eerie power I thought I had discovered when I could see through my hand. You can take this one step further by doing as John Barth did with his "Frame-Tale": on one side of a thin page, write, "ONCE UPON A TIME," and on the other side, "THERE WAS A STORY THAT BEGAN," and then tape the ends together with a twist. Start reading the story as you trace the curve of the Möbius strip. You find that the story never ends. That's the exact feeling that recursion instills in folks who appreciate it—it's a simple-looking loop, but with a literal twist. And with just that twist, it enters a different world.

When it comes to writing recursion into computer programs, it's as simple as defining an idea as directly related to the idea itself. When first learning how to draw a tree, you see that this is reflected in nature. A tree starts with a vertical line that has a few lines popping out of its top. To proceed to draw the tree further, you take each of the lines at the top and repeat the act of adding more lines to pop out of each top. And so forth. In the end you get a tree with a lot of subbranches created by simply using the same method you started with. In other words, a tree branch is composed of tree branches. The part is defined by the part itself. You see this played out in the exact opposite direction from the sky into the ground underneath a tree with its root system—so nature paints with recursion quietly and obviously.

Another way to consider the magic of recursion is to look at a proprietary operating system called Unix that renegade programmers attempted to completely rewrite from scratch in the 1980s. Rather than let Unix be controlled by its owner, AT&T, MIT's

Richard Stallman wanted such an important software system to be free of any constraints. He called his endeavor the GNU Project. The name GNU epitomizes the idea of recursion, as it signifies "(G)NU's (N)ot (U)nix." Pause for a moment as you read that sentence. So the "U" makes sense as standing for "(U)nix," and the "N" makes sense as standing for "(N)ot." It's when you get to the "G" as standing for "(G)NU" that things get a bit weird. If we try to expand out the "G" and spell it out successively, you can see the infinite nature of this expression:

```
[G]NU's Not Unix
[[G]NU's Not Unix]NU's Not Unix
[[[G]NU's Not Unix]NU's Not Unix]NU's Not Unix
[[[[G]NU's Not Unix]NU's Not Unix]NU's Not Unix]NU's
    Not Unix
[[[[[G]NU's Not Unix]NU's Not Unix]NU's Not Unix]NU's
    Not Unix]NU's Not Unix

    . . .
```

Relatedly, a physical metaphor that comes close is a Russian matryoshka doll—a doll that has a smaller but identical version of itself nested inside it, and on and on. That's because we can say that a matryoshka doll is made from another matryoshka doll, and so on and so forth. But even with the world record matryoshka doll set, you will run out of dolls by the time you pull out the fifty-first one; when it comes to computational matryoshka dolls, there are no specific limits to how deeply they can be nested within each other unless a "base case" is set by the programmer explicitly. Imagine opening a doll after smaller doll after smaller doll after even a doll that is the size of a rice kernel, and you can still continue to find another doll inside the doll inside the doll.

On a more practical level beyond playing with dolls, mathematically minded folks can take the idea of recursion and create elegant expressions of concepts that look nothing like the messier tofu shop accounting program that I showed you before. Recursion differs from the brute force expression of a loop which likens itself to more like a conveyor belt on an assembly line. You lay out all the tasks, and then you tell each task to perform in sequence. And then you GOTO the beginning of the list and do it over again, like on an assembly line. Recursion is stylistically different in nature where you define the task to be performed in terms of itself—like laying out the steps to make a large pot of curry on an assembly line where one key ingredient is a smaller pot of curry. You end up with an assembly line that vanishes inside the process of making the smaller pot of curry, which in turn will require a smaller pot of curry, and so forth—you simply disappear inside the thing you're creating. It's not a concept for the fainthearted. The central idea is to express the definition of something with a definition itself, which is a vaguely imaginable idea that doesn't have a home in the real world but is completely native in the realm of cyberspace.

So now you know there are certain forms of elegantly expressing oneself in the computational world that are akin to the question-inducing power of art as we know it in our physical world. Along those lines, computational thinkers have appreciation for a kind of highly conceptual art that isn't yet at the Metropolitan Museum of Fine Art. When programmers say "code is poetry," they really mean it. Recursion is an unusually compact way to express complex ideas that can be infinite in nature and are deeply paradoxical, like what happens when you try to unpack "GNU." In computation, it becomes possible to build an enigma into actual working machinery, but even before the computer era, recursion was a captivating philosophical concept. Or, expressed succinctly by Michael Corballis in his entire book on the topic of recursion from a humanist's perspective:

Recursion (rĭ-kûr'-zhən) *noun*. If you still don't get it, see **recursion**.

5. Loops are indestructible unless a programmer has made an error.

Think back to how we got started in this chapter with a loop that let us count to one billion:

```
top = 1000000000
i = 0
while i < top: i = i + 1
```

I just timed this running on my computer and it completed in under a minute. Keep in mind that by the time this book is printed and you have it in your hands, computing machines will have become even faster. The counter runs unimpeded—much like if you were behind the wheel of a fast car on a road that extended forever and hit the gas pedal. But what if there were a big rock a few miles down the road that—with your stereo blasting and your adrenaline on fire—you could easily fail to notice while speeding? That's right. *Blam!* Your car will likely wipe out, and hopefully you'll have your seat belt on.

There's a difference I'd like you to consider when thinking about the counting loop above as compared with my car analogy. The car will start to accelerate and at some point hit top speed. As it encounters the rock you will violently experience an "ouch" moment for at least a few seconds. You'll have time to regain your senses and then step out of the fire and debris, hopefully with only a few minor flesh wounds.

The computational process, once initiated, will run at top speed from the very moment it comes alive. And if it were to hit some kind of error, it would immediately stop. The entire world it lived within would vanish in that same instant too. In the instant when the computing machine has been involuntarily stopped, it's an absolute catastrophe. Because when computation is doing its thing, you can't see it doing its thing. But when it's not working, it will either complain to you explicitly or it will simply freeze. I'm sure you've seen this happen before. Some message is flashed on your screen or the computer screen simply goes blank. You've usually had no warning at all before it's about to happen—and it usually makes you a bit unhappy, or even angry. Just search for "computer rage" on the internet and you'll feel in good company.

Before we go into why computers crash, let's consider what it feels like for the computer. The best analogy I can think of involves the many "epic" domino setup experts you see online who painstakingly lay out thousands of domino tiles and then turn on the video camera to watch the dominos fall in perfect sequence until one tile has been placed incorrectly and . . . FAIL. The embarrassment and shock are real, and there's only one recourse: go back and fix everything, right from the very start.

It takes just one misplaced domino to destroy hundreds of items moving in perfect sequence together—this is what it's like when the software application you're running comes to a complete halt. And it's with the same straightforward attitude, with all the little dominos strewn everywhere across the floor, that a programmer needs to gleefully say: "It needs to be fixed." Much like the expert domino placer's disciplined patience to fix and redo everything, a programmer must behave in exactly the same manner. If computer programmers became uncontrollably angry each time a piece of software crashed, they wouldn't get any work done. Because software crashes a lot, you'll tend to find that people who write software professionally have an unusually high tolerance for catastrophes while also having little tolerance for minor mistakes that could easily be avoided.

Imagine a job where every few minutes you're likely to be told, by a computer, that you did something wrong. The more complex the program or computing system upon which it is running, the more things that can go wrong. These fall into three categories: avoidable ("dumb") errors, less avoidable errors, and unavoidable errors. In the early days of computing, many software systems and the hardware on which they ran had kinks in them because they were like experimental aircrafts—so many errors were in the "unavoidable" category. But that's less often the case these days, because computing machines have matured exceedingly well. For example, when I was writing computer programs in the eighties, it wasn't uncommon to run into errors that were out of my control due to the "laboratory testing" nature of some of those early machines. It's quite revealing that the term for these errors, or "bugs," originated in Grace Hopper's 1940s discovery of a moth trapped inside a relay inside an early computing machine—electricity couldn't make it across two contact points, preventing the computer from working properly.

The fact that Hopper literally found a computer bug by wading through the complex wiring of machines back in the day is symbolic: it's quite easy to imagine her joy after physically removing the pesky moth. It's also easy to imagine the tortured search for the problem, which can often feel like looking for a needle in a haystack of numbers and symbols in a computer program. Since it's so hard to find bugs in software, there's a strong bias for programmers working collaboratively on a project to stigmatize the easily avoidable "dumb" errors within their team. Fortunately, today there are a variety of systems and technologies that cut down on software bugs, but it's simply humanly impossible to make bug-free software. And it's important to remember that not all bugs are fatal—many kinds of bugs can live within the software and have no obvious impact on operations.

An easy, but also uneasy, way to think of why it's impossible to build a bug-free computer program can be felt while you're flying in an airplane. Considering it's made of millions of parts, think about how many screws might be loose or missing. Likewise, consider how a complex piece of software like Microsoft Excel will have tens of millions of lines of computer code—what's the likelihood that one line might have been typed or conceived incorrectly? Keep in mind that a major software program could have been built by hundreds or thousands of people over many years. One loose screw is not likely to be fatal to an airplane's operation in the same way that one erroneous number setting inside Excel's code will not necessarily bring it to a halt. Yet you can imagine that if enough benign bugs are littered sloppily all about, they can start to have a negative impact on each other and unexpected things might happen.

Computing machines can loop in unflagging and untiring ways. An expansive, hidden universe is accessed when a computer program comes alive and its loops start to clock through instructions. Digits are freely read and written with precision and without friction within a purely numerical world. Depending upon the level of complexity of the software, there will always be some likelihood that a bug lies waiting to impact the program's execution and possibly bring it to a full stop. When it halts, if there's a software person who can find the bug and make the necessary repair, you'll be back and looping away. The harder things to find are the bugs that don't immediately present themselves as problematic but are quietly rattling away to potentially cause trouble in ways that you will have difficulty diagnosing.

Don't forget that what you see displayed on a computer screen is only a tiny fraction of what's actually happening within the invisible world of computing machinery. There are endless streams of digital information that are hurriedly getting processed, both elegantly and inelegantly, behind the scenes powered by loops, and some recursive loops too. We have human computers, human hardware makers, and human software engineers to thank for this. And always be ready for the impact of an occasional invisible moth left by a fellow human being—at the most inopportune moment. Be prepared, however, for a future where computers can reach inside themselves to relentlessly remove the bugs we put inside them—removing any and all obstacles to make their powerful looping activities interminable. Repeating themselves unerringly, forever.

Chapter 2

MACHINES GET
LARGE

1 // Embracing exponential thinking is unnatural at first.

2 // Loops wrapped inside loops open new dimensions.

3 // Be open to both directions of the powers of ten.

4 // Losing touch with human scale can make you toxic.

5 // Computers team up with each other way better than we do.

1. Embracing exponential thinking is unnatural at first.

There's an old riddle about a beautiful, pristine pond that has become invaded by an exotic species of lily pads. This breed of lily pads doubles in number each day, quickly spreading to cover the entire surface of the pond. A scientist lives in a little house next to the pond, and she carefully tracks the invasive plant species growth each day with particular concern for the water life that is becoming choked of sunlight in the shadow of the lily pads.

By the thirtieth day of the invasion, the pond is completely covered, with a total of 536,870,912 lily pads. So the riddle goes: "On the fifteenth day, roughly how many lily pads covered the pond?" I know when I first heard this riddle, I thought it was going to be half of 536,870,912, or 268,435,456 lily pads. Wrong.

It's easier to come to the answer when asked, "On what day is the pond half covered with lily pads?" The answer is the twenty-ninth day. Why? Because the lily pads double in number each day, so it's logical that the day prior to the thirtieth is when the pond was half covered. In case you got that wrong, it's fairly natural to want to think it was half covered on the fifteenth day, since that is halfway through the thirty days.

Let's think about what really happens by the fifteenth day if we double the number of lily pads, starting with one lily pad on the first day.

```
Day 1 = 1 lily pad
Day 2 = 2 lily pads
```

Day 3 = 4 lily pads
Day 4 = 8 lily pads
Day 5 = 16 lily pads
Day 6 = 32 lily pads
Day 7 = 64 lily pads
Day 8 = 128 lily pads
Day 9 = 256 lily pads
Day 10 = 512 lily pads
Day 11 = 1,024 lily pads
Day 12 = 2,048 lily pads
Day 13 = 4,096 lily pads
Day 14 = 8,192 lily pads
Day 15 = 16,384 lily pads

Given that the Excel formula looks like = POWER(2, DAY-1) we can verify the number of lily pads on the thirtieth day as:

Day 30 = POWER(2, 30-1) = POWER(2, 29) = 536,870,912
    lily pads

On the fifteenth day, we're at 16,384 lily pads, so that's still only 0.003 percent of the way to 536,870,912 lily pads—nowhere near the 50 percent mark. Running the formula further, past the fifteenth day, we find that we finally break 1 percent of the total number (5 million lily pads) between the twenty-third and twenty-fourth day:

Day 23 = POWER(2, 22) = 4,194,304
Day 24 = POWER(2, 23) = 8,388,608

If all this feels a bit difficult to conceptualize, it's because it illustrates the difference between exponential thinking and linear thinking—linear thinking is how we're wired.

When people mistakenly answer the riddle by saying the pond is 50 percent covered on the fifteenth day, they're using linear thinking. We're used to working with linear thinking because it's how we know to make sense of our world. Think of how a seedling you plant, then water daily, is going to grow gradually each day. The only thing that might make it grow faster would be some fertilizer. But if left alone, it's going to grow in an incremental manner. Steady, linear growth of course doesn't have to mean slow growth—for instance, when children learn how to count by tens instead of ones, they get excited when they blaze past a hundred. But even when you count by millions, you're still living in the commons of linear thinking by taking even steps of equal amounts.

Exponential thinking, as illustrated by the actual answer to the riddle, is what you become accustomed to in the computational world, not only because it governs the doubling power of Moore's law, but also because of the way that loops are often crafted. I like to think of the difference between linear thinking and exponential thinking as akin to the difference between additive effects and multiplicative effects. Addition makes a number get bigger by a set increment; multiplication makes a number bigger by a set "leap." For example, when I count by tens like an eager elementary schooler who has just discovered this superpower, I start with 1, add 10 each time, and in just ten steps I can leap to 101. But if I instead multiply by tens starting from 1, I'm already at 10 billion after ten iterations—which feels akin to a super-

superpower for any budding mathlete that discovers this twist. We're using the same old boring number 10 in both instances, but just rotating that plus sign (+) by 45 degrees into a times symbol (×) puts us in a whole different dimension. That's because multiplication "hides" a bunch of additions inside itself.

Consider how 5 × 10 means taking the number 5 and adding it to itself a total of nine times: 5 + 5 + 5 + 5 + 5 + 5 + 5 + 5 + 5. Or consider taking this further and calculating 5 × 1,000. It would look like this:

5 + 5 + 5 + 5 + 5 + 5 + 5 + 5 + 5 + 5 + 5 + 5 + 5 + 5 + 5 + 5 + 5 +
5 + 5 + 5 + 5 + 5 + 5 + 5 + 5 + 5 + 5 + 5 + 5 + 5 + 5 + 5 + 5 + 5 +
5 + 5 + 5 + 5 + 5 + 5 + 5 + 5 + 5 + 5 + 5 + 5 + 5 + 5 + 5 + 5 + 5 +
5 + 5 + 5 + 5 + 5 + 5 + 5 + 5 + 5 + 5 + 5 + 5 + 5 + 5 + 5 + 5 + 5 +
5 + 5 + 5 + 5 + 5 + 5 + 5 + 5 + 5 + 5 + 5 + 5 + 5 + 5 + 5 + 5 + 5 +
5 + 5 + 5 + 5 + 5 + 5 + 5 + 5 + 5 + 5 + 5 + 5 + 5 + 5 + 5 + 5 + 5 +
5 + 5 + 5 + 5 + 5 + 5 + 5 + 5 + 5 + 5 + 5 + 5 + 5 + 5 + 5 + 5 + 5 +
5 + 5 + 5 + 5 + 5 + 5 + 5 + 5 + 5 + 5 + 5 + 5 + 5 + 5 + 5 + 5 + 5 +
5 + 5 + 5 + 5 + 5 + 5 + 5 + 5 + 5 + 5 + 5 + 5 + 5 + 5 + 5 + 5 + 5 +
5 + 5 + 5 + 5 + 5 + 5 + 5 + 5 + 5 + 5 + 5 + 5 + 5 + 5 + 5 + 5 + 5 +
5 + 5 + 5 + 5 + 5 + 5 + 5 + 5 + 5 + 5 + 5 + 5 + 5 + 5 + 5 + 5 + 5 +
5 + 5 + 5 + 5 + 5 + 5 + 5 + 5 + 5 + 5 + 5 + 5 + 5 + 5 + 5 + 5 + 5 +
5 + 5 + 5 + 5 + 5 + 5 + 5 + 5 + 5 + 5 + 5 + 5 + 5 + 5 + 5 + 5 + 5 +
5 + 5 + 5 + 5 + 5 + 5 + 5 + 5 + 5 + 5 + 5 + 5 + 5 + 5 + 5 + 5 + 5 +
5 + 5 + 5 + 5 + 5 + 5 + 5 + 5 + 5 + 5 + 5 + 5 + 5 + 5 + 5 + 5 + 5 +
5 + 5 + 5 + 5 + 5 + 5 + 5 + 5 + 5 + 5 + 5 + 5 + 5 + 5 + 5 + 5 + 5 +
5 + 5 + 5 + 5 + 5 + 5 + 5 + 5 + 5 + 5 + 5 + 5 + 5 + 5 + 5 + 5 + 5 +
5 + 5 + 5 + 5 + 5 + 5 + 5 + 5 + 5 + 5 + 5 + 5 + 5 + 5 + 5 + 5 + 5 +

5+5+5+5+5+5+5+5+5+5+5+5+5+5+5+5+5+5+5+
5+5+5+5+5+5+5+5+5+5+5+5+5+5+5+5+5+5+5+
5+5+5+5+5+5+5+5+5+5+5+5+5+5+5+5+5+5+5+
5+5+5+5+5+5+5+5+5+5+5+5+5+5+5+5+5+5+5+
5+5+5+5+5+5+5+5+5+5+5+5+5+5+5+5+5+5+5+
5+5+5+5+5+5+5+5+5+5+5+5+5+5+5+5+5+5+5+
5+5+5+5+5+5+5+5+5+5+5+5+5+5+5+5+5+5+5+
5+5+5+5+5+5+5+5+5+5+5+5+5+5+5+5+5+5+5+
5+5+5+5+5+5+5+5+5+5+5+5+5+5+5+5+5+5+5+
5+5+5+5+5+5+5+5+5+5+5+5+5+5+5+5+5+5+5+
5+5+5+5+5+5+5+5+5+5+5+5+5+5+5+5+5+5+5+
5+5+5+5+5+5+5+5+5+5+5+5+5+5+5+5+5+5+5+
5+5+5+5+5+5+5+5+5+5+5+5+5+5+5+5+5+5+5+
5+5+5+5+5+5+5+5+5+5+5+5+5+5+5+5+5+5+5+
5+5+5+5+5+5+5+5+5+5+5+5+5+5+5+5+5+5+5+
5+5+5+5+5+5+5+5+5+5+5+5+5+5+5+5+5+5+5+
5+5+5+5+5+5+5+5+5+5+5+5+5+5+5+5+5+5+5+
5+5+5+5+5+5+5+5+5+5+5+5+5+5+5+5+5+5+5+
5+5+5+5+5+5+5+5+5+5+5+5+5+5+5+5+5+5+5+
5+5+5+5+5+5+5+5+5+5+5+5+5+5+5+5+5+5+5+
5+5+5+5+5+5+5+5+5+5+5+5+5+5+5+5+5+5+5+
5+5+5+5+5+5+5+5+5+5+5+5+5+5+5+5+5+5+5+
5+5+5+5+5+5+5+5+5+5+5+5+5+5+5+5+5+5+5+
5+5+5+5+5+5+5+5+5+5+5+5+5+5+5+5+5+5+5+
5+5+5+5+5+5+5+5+5+5+5+5+5+5+5+5+5+5+5+
5+5+5+5+5+5+5+5+5+5+5+5+5+5+5+5+5+5+5+
5+5+5+5+5+5+5+5+5+5+5+5+5+5+5+5+5+5+5+
5+5+5+5+5+5+5+5+5+5+5+5+5+5+5+5+5+5+5+
5+5+5+5+5+5+5+5+5+5+5+5+5+5+5+5+5+5+5+
5+5+5+5+5+5+5+5+5+5+5+5+5+5+5+5+5+5+5+

5 + 5 + 5 + 5 + 5 + 5 + 5 + 5 + 5 + 5 + 5 + 5 + 5 + 5 + 5 + 5 +
5 + 5 + 5 + 5 + 5 + 5 + 5 + 5 + 5 + 5 + 5 + 5 + 5 + 5 + 5 + 5 +
5 + 5 + 5 + 5 + 5 + 5 + 5 + 5 + 5 + 5 + 5 + 5 + 5 + 5 + 5 + 5 +
5 + 5 + 5 + 5 + 5 + 5 + 5 + 5 + 5 + 5 + 5 + 5 + 5 + 5 + 5 + 5 +
5 + 5 + 5 + 5 + 5 + 5 + 5 + 5 + 5 + 5 + 5 + 5 + 5 + 5 + 5 + 5 +
5 + 5 + 5 + 5 + 5 + 5 + 5 + 5 + 5 + 5 + 5 + 5 + 5 + 5 + 5 + 5 +
5 + 5 + 5 + 5 + 5 + 5 + 5 + 5 + 5 + 5 + 5 + 5 + 5 + 5 + 5 + 5 +
5 + 5 + 5 + 5 + 5 + 5 + 5 + 5 + 5 + 5 + 5 + 5 + 5 + 5 + 5 + 5 +
5 + 5 + 5 + 5 + 5 + 5 + 5 + 5 + 5 + 5 + 5 + 5 + 5 + 5 + 5 + 5 +
5 + 5 + 5 + 5 + 5 + 5 + 5 + 5 + 5.

So multiplying a number by 1,000 packs a punch and grows the number 5 at Jack and the Beanstalk speed, where one tiny bean grows into a giant stalk extending into the sky the morning right after it's planted. To further cement this point, let your mind scan the "+ 5s" above and feel the incredible distance afforded by a multiplicative "leap" whereby you are shown the hidden power embodied in the × that a regular + doesn't have.

Exponential growth is native to how the computer works. This is how the amount of computing memory available has evolved. The same can be said about processing power. So when you hear people in Silicon Valley talk about the future, it's important to remember that they're not talking about a future that is *incrementally* different year after year. They're constantly on the lookout for *exponential* leaps—knowing exactly how to take advantage of them because of their fluency in speaking machine. Let's next dip into how something so seemingly boring as speaking loops can make exponential sorcery happen within the computer.

2. Loops wrapped inside loops open new dimensions.

Recall how you first learned to draw a three-dimensional cube on paper back in elementary school. You start with two squares slightly offset. Then you connect the corners of each square to the other. The plane of the paper is flat, but your brain sees the line drawing of the cube and can't help but imagine that there's more space now available to you on the page. At first that seems impossible, because there's been no net addition of any new space and all you have is the optical illusion of it. Consider, though, how you technically do have more space available to you with the cube as drawn—because now you can place points within the cube and tap into an entirely new dimension. For this to "click" for you, you need to activate your imagination a bit: consider how before you drew the cube you didn't have access to a three-dimensional space of any size, and depending on how you draw it you can make it bigger than the plane of paper itself.

Let's do this again using a different formulation that mathematicians use as a kind of gateway to the fourth dimension. Start off by drawing a point. This signifies zero dimensions.

•

Then extend that point in space and connect the two points. You get a line. This signifies one dimension.

Then extend that line in space and connect the four points. You get a plane. This signifies two dimensions.

Then extend that plane in space and connect the eight points. You get the cube you already know. It signifies three dimensions.

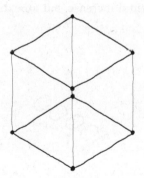

So how do you think you can depict the fourth dimension? Right! Take the cube, extend it in space, and connect the sixteen points. You get a hypercube, which signifies four dimensions.

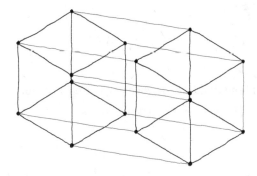

With each dimensional addition, I want you to feel what is a typical exponential shift. For example, when we move from one dimension to two dimensions with, say, a 10-millimeter line projected to a 10-millimeter square, our new space coverage is 100 square millimeters. That's a big jump in amount of space, and when we move to three dimensions that's an even larger space—and it isn't merely an incremental increase, but an extradimensional increase.

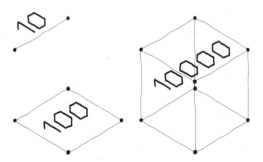

We move from 100 square millimeters to covering 10,000 cubic millimeters in three dimensions. With each dimensional shift, our growth in space is literally exponential. I know this is all pretty abstract, and although I would love to give you an easy-to-digest physical metaphor, I only have an invisible one: loops. Are you ready?

Let's start by considering how a decade is ten years, and how a year is divided into twelve months. We traverse a decade by looping through ten years:

```
for( year = 1; year <= 10; year = year+1 ) { }
```

This piece of code starts at year = 1, increments itself by 1, and terminates when year is greater than 10. It currently does nothing important because it's applied to a block of code that has nothing inside it. The { } can be thought of as giving a tight, squeezing "hug" so that the code inside sticks together as if it were one entity. Count it out.

1. 2. 3. 4. 5. 6. 7. 8. 9. 10.

Next, let's consider looping over the twelve months in a year by reusing the first snippet of code with "year" replaced with "month" and stopping at 12 instead of 10:

```
for( month = 1; month <= 12; month = month+1 ) { }
```

Count it out.

1. 2. 3. 4. 5. 6. 7. 8. 9. 10. 11. 12.

Now let's put the month loop into the year loop.

```
for( year = 1; year <= 10; year = year+1 ) {
    for( month = 1; month <= 12; month = month+1 ) { }
}
```

What happens is the years are stepped over for ten times, and for each year the months are stepped over for twelve times too. If you were to count it out, it would look something like:

1.
1. 2. 3. 4. 5. 6. 7. 8. 9. 10. 11. 12.
2.
1. 2. 3. 4. 5. 6. 7. 8. 9. 10. 11. 12.
3.
1. 2. 3. 4. 5. 6. 7. 8. 9. 10. 11. 12.
4.
1. 2. 3. 4. 5. 6. 7. 8. 9. 10. 11. 12.
5.
1. 2. 3. 4. 5. 6. 7. 8. 9. 10. 11. 12.
6.
1. 2. 3. 4. 5. 6. 7. 8. 9. 10. 11. 12.
7.
1. 2. 3. 4. 5. 6. 7. 8. 9. 10. 11. 12.
8.
1. 2. 3. 4. 5. 6. 7. 8. 9. 10. 11. 12.
9.
1. 2. 3. 4. 5. 6. 7. 8. 9. 10. 11. 12.
10.
1. 2. 3. 4. 5. 6. 7. 8. 9. 10. 11. 12.

Did you notice how that felt? By simply placing one loop inside another loop, something happened that shouldn't feel natural at all to you. Now, if you were to assume that each month contained exactly thirty days—to make this example easier to understand— then you'd use a piece of code like:

```
for( day = 1; day <= 30; day = day+1 ) { }
```

Let's count it out:

1. 2. 3. 4. 5. 6. 7. 8. 9. 10. 11. 12. 13. 14. 15. 16. 17. 18. 19. 20. 21. 22. 23. 24. 25. 26. 27. 28. 29. 30.

Can you imagine what will happen if you place this loop inside the innermost loop that walks over months?

```
for( year = 1; year <= 10; year = year+1 ) {
  for( month = 1; month <= 12; month = month+1 ) {
    for( day = 1; day <= 30; day = day+1 ) { }
  }
}
```

I prefer not to waste all the paper that would go into tracing out this case, but if you want to do so yourself, you'd start off with:

1.
1.

1. 2. 3. 4. 5. 6. 7. 8. 9. 10. 11. 12. 13. 14. 15. 16. 17. 18. 19. 20. 21. 22. 23. 24. 25. 26. 27. 28. 29. 30.

2.

1. 2. 3. 4. 5. 6. 7. 8. 9. 10. 11. 12. 13. 14. 15. 16. 17. 18. 19. 20. 21. 22. 23. 24. 25. 26. 27. 28. 29. 30.

3.

1. 2. 3. 4. 5. 6. 7. 8. 9. 10. 11. 12. 13. 14. 15. 16. 17. 18. 19. 20. 21. 22. 23. 24. 25. 26. 27. 28. 29. 30.

. . .

and just keep going through the rest of the 3,600 numbers that you've generated from working in three dimensions by tracing ten years, with twelve months apiece, each containing thirty days. That's a lot like taking your line, extending it out as a plane, and then extending that plane out as a rectangular solid. What all this means in practical terms is that when loops go inside of loops, it's like an electric spark that triggers a turbo boost for whatever gets dropped into the lovingly hugged { block }.

Anything can be dropped into a { block }. Say, for instance, you've rented ten computing machines on the network somewhere. It would be easy to get all ten of them to do your task of stepping through ten years by looping over all your available machines.

```
for( machine = 1; machine <= 10; machine = machine+1 ) {
    for( year = 1; year <= 10; year = year+1 ) {
        for( month = 1; month <= 12; month = month+1 ) {
```

```
        for( day = 1; day <= 30; day = day+1 ) { }
    }
  }
}
```

Or else you could flip that logic and instead do something on all ten machines for each day counted over a ten-year span:

```
for( year = 1; year <= 10; year = year+1 ) {
    for( month = 1; month <= 12; month = month+1 ) {
        for( day = 1; day <= 30; day = day+1 ) {
            for( machine = 1; machine <=10; machine =
            machine+1 ) { }
        }
    }
}
```

A few thoughts might come to mind:

- There's nothing stopping you from changing the limit of year <= 10 to year <= 100,000.
- You can continue counting all the way down to the hours, minutes, and seconds this way.
- If we rented a few thousand machines, it would be as easy as changing machine <= 10 to as many machines that you can access.

Each successive loop introduces a new dimension, much in the same way that we made a point into a line, and a line into a plane, and then a plane into a cube. With each successively hugged or "nested" loop, another dimension of possibilities opened up. Space that didn't exist prior to nesting a loop suddenly existed, and tweaking the start and end limits of each dimension made the additional space bigger or smaller. In short, it's a means to open up spaces that are much larger than the ones that sit in front of us or surround us at the physical scale of our neighborhoods or entire cities. There are literally no limits to how far each dimension can extend, and no limits to how many dimensions can be conjured up with further nesting of loops. This should feel unnatural to those of us who live in the analog world, but it's just another day inside the computational universe.

3. Be open to both directions of the powers of ten.

One of the best ways to translate the powerful feeling of transcending our limited perspective of the spaces around and inside us is to view the short film *Powers of Ten* by the late designers Ray and Charles Eames. The Eames duo are better known for their expensive chairs that you can find in many a fancy home, but their lesser-known film work gives the clearest glimpse of what was really going on in their dynamic minds. *Powers of Ten* is available to view online for free, and I've found that it's the fastest way to understand the disturbing yet powerful feeling of omniscience that emerges when you work deeply in the computational medium.

The nine-minute film begins with a close-up overhead shot just one meter above the ground aimed at a couple napping on a picnic blanket in a Chicago park. The camera then begins to zoom out by one power of ten to get to 10 meters with the couple getting relatively smaller, and then to 100 meters with the entire park visible from the air, and then to 10,000,000 meters with the whole earth in view, and then to 10,000 million meters as we pass the orbit of Venus, and then we keep going all the way out past our galaxy to 10,000 light-years. We then zoom in all the way back to 1 meter away from the couple in the park and then down to 10 centimeters, where we can see the man's skin up close, and then to 1 millimeter, where we start to enter the pores of his skin, and then all the way down to 0.00001 angstroms (one angstrom is one ten-billionth of a meter). It's more compelling when seen animated, but you can imagine how it aptly illustrates both the grand scale of the universe and the miniscule scale at which atoms exist.

With your imagination fluidly tracing every zoomed-in or zoomed-out increment, you can feel like you've transcended the operating scale of your default existence. If you were born before the year 2000, when the pinch-to-zoom gesture on touch screens was still new, you might still remember the feeling of "wow" when, just with your fingertips, you could intuitively summon an extra magnification power of ten. But unlike the film *Powers of Ten*, there's a limit you quickly run into as you zoom in or out of a picture, because the subresolution of your screen pixels starts to run out. At some point, all you can see are the underlying rectangular pixels instead of the micro-crinkly surface of someone's face. In addition, when you zoom out of a photo, you eventually reach the edges of the frame, at which point there's no reason to keep zooming out.

I know these limits well because much of my time as a visual designer in the early nineties involved crafting images that pushed beyond the edges of a normal photograph at the scale of buildings. Meanwhile, I paid attention to extremely fine details that could be rendered only with special computerized printing methods that required a magnifying glass to appreciate. A set of those experiments, the *Morisawa 10* posters, made it into the permanent collection of the Museum of Modern Art (MoMA)—which you can likely recognize are simply the product of loops and more loops with repeating elements and sub-elements.

 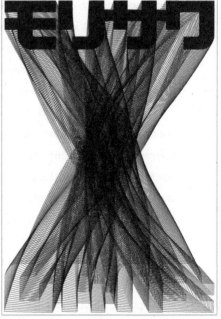

But all of those works were in print, so at some point they had to end at the edges of the paper on which they were printed. Also, a print is ultimately made out of ink dots, so when you zoom into the image you'll end up at the dots. A more appropriate means to represent computational thinking at the infinite scale is to graduate from the surface of a printed page and to examine the power of recursion, and to go beyond the twist of the Möbius strip. Let's take the famous mathematical example of a class of geometric figures called "fractals," which demonstrates recursion. We start by considering a set of four lines arranged as such:

Go ahead and draw this in pencil on a surface near you. Now replace each of the line segments (all four of them) with an exact replica of that sequence of four shorter lines you originally drew. You can see it quickly becomes jagged and spiky, like the back of a seahorse:

This is called a Koch curve, named after Swedish mathematician Niels Fabian Helge von Koch, and dates to the turn of the last century. When you take three equisized Koch curves and attach them end to end to form a triangle at the endpoints, you get a six-sided star. Go ahead and draw it on paper to verify for yourself. This is what it will look like:

It starts off as a star, but when you replace each line with a Koch curve, you get what's called a Koch snowflake, for obvious reasons as you can see below. This becomes immediately apparent as you successively continue each new cycle of line replacement. The details will get finer and finer as you progress—no matter how close you zoom into a Koch snowflake. The details are just repeated, of course, and they're repeated forever. You ask yourself, *When does it stop?* And the answer is, simply, *Never.*

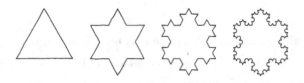

A fun, truly magical aspect of the Koch snowflake is that its perimeter is infinite but its area is finite. The former makes sense, because you can imagine replacing every _ with _/_, increasing the length of the outline. But the actual area covered by this snowflake is mathematically proven to hit a limit, even though its perimeter keeps on growing. This makes no sense in the physical world. It's as if I told you there was a pond that you can cover with an exact number of lily pads, but it will take you an infinite amount of time to circumnavigate. But such wonders are par for the course in the unusual world of computation.

Computation has a unique affinity for infinity, and for things that can be allowed to continue forever, which take our normal ideas of scale—big or small—and easily mess with our mind. You may think you are not in control, but you are in complete control when you write the codes and construct the loops to your liking. There is a certain comfort as you come to realize that, with eventual ease, you can craft infinitely large systems with also infinitely fine details. It's trivial to define a computational means to traverse across a billion users and analyze their individual minutiae using a couple of well-crafted loops. It will look something vaguely like:

```
for( user = 1; user <= 1000000000; user = user+1 ) {
    user_data = get_data(user);
        for( data = 1; data <= length(user_data); data =
            data+1 ) {
        analyze_user_data(data);
    }
}
```

where we inspect every individual data item in a billion users, which could correspond to photos, keystrokes, or GPS locations. There's no real upper bound, nor is there any limit to how fine-grained any computational process can be implemented by whoever constructs the codes. There's no need to choose "how broad" or "how deep," because the answer can simply be "both." And if there's any concern that the code might take too long to run, just wait for a few cycles of Moore's law to tick away, and you can expect to eventually have enough processing power in just a few more years.

So think of this all as a kind of superpower where you can jump as high into outer space as you desire to gain vast and broad perspective by scaling the positive powers of ten: 1, 10, 100, 1,000, 10,000, 100,000, 1,000,000, 10,000,000, 100,000,000, and even larger. You also possess the flip opposite of this superpower, where you can shrink as small as you need to gain fine and precise perspective by scaling the negative powers of ten: 1, 0.1, 0.01, 0.001, 0.0001, 0.00001, 0.000001, 0.0000001, 0.00000001, 0.000000001, 1 angstrom, and even smaller. Pretty cool, huh? Unnatural, right? Almost . . . alien. You got it. *Cue Bowie's "Life on Mars" . . .*

4. Losing touch with human scale can make you toxic.

"Complicated" means something that is knowable, and although it may take time, it's wholly possible to understand. You might just need some good old-fashioned brute force to do it—you'll get tired in the process, but it's doable. A complicated machine (think of the printing press or digital delivery service that brought you this book) is understandable.

"Complex" means something that is not knowable, and even brute force can't easily tackle it. I say easily because in the twenty-first century the computing power we have access to is absurdly amazing, but it still can't solve complex problems—yet. A complex machine (think of any human being you have a relationship with) is not understandable.

I always find it necessary to keep this distinction in the foreground, because how we make systems out of computation is generally complicated, but how we humans relate to the computational systems we make has complex effects that we're still figuring out. If you think back to the *Powers of Ten* and loops within loops and the infinite depth of fractals, and the ability to control time and space at the scale of infinity within the precision of fractions of an angstrom, you can see how a person who lives and breathes this invisible world of absolute power and who controls every minute of the day might lose their connection with reality. Or, as the late artificial intelligence (AI) pioneer Joseph Weizenbaum presciently wrote:

> The computer programmer is a creator of universes for which they alone are the lawgiver. No playwright, no stage director, no emperor, however powerful, has ever exercised such absolute authority to arrange a stage or field of battle and to command such unswervingly dutiful actors or troops.

For that reason, I don't find it surprising at all when folks who write code start to develop an unusual relationship with reality, or even veer into the zone of becoming slightly mad, considering the level of power and control that coding gives them. Playing a video game might seem to grant an analogous amount of power to a gamer—but writing the video game code itself kicks the power way up to being akin to godlike status. I'm not saying that every coder ends up with some kind of god complex. However, when other destructive personality traits get mixed in, the result is the archetypal "brogrammer"—a male coder who can impose his delusions and sense of entitlement on others to a terribly uncomfortable level. Keep in mind that coding does not causally create this outcome and is only correlative—there are plenty of friendly programmers out there. And even absent the fluency of coding, we can see a related behavior when despots and other powermongers use social media to impact millions of minds, and their moods, with just a few destructive keystrokes from anywhere on earth.

The actual machinery of computing is complicated yet understandable; the social impact of this complicated machinery becomes complex when it involves as many humans as it does today. Complicated situations are ultimately resolvable, but complex situations are entirely different—though we should still try to understand them because they impact ourselves and our fellow human beings.

Wielding billions of bits and traveling at speeds of millions of cycles per second, with everything moving, behaving, working according to your exact instructions, with no complaints or dissent, is the definition of dictatorial control. So it's not hard to imagine why some coders who have been operating gigantic systems at once unimaginable scales (but now imaginable to you) for a few decades might start feeling a bit powerful and trivialize the human beings around them who do not obey their keystroked commands. When I look at the snarky, annoyed, nasty behavior online by some of the technologically fluent men who literally grew up in cyberspace, I partially attribute it to the fact that coding can become extremely intoxicating—so it's not a surprise to me when I occasionally encounter discomforting behavior among the most technically fluent. Keep in mind that composing lines of code is an extremely creative task that intrinsically involves generosity through sharing skills with others—it does not make you a bad person. But it definitely can change your perspective on the world around you if you aren't careful.

I'm the first to admit that, in my early years of personal development, coding negatively affected me and all of my relationships. In my midforties, I was fortunate to have a close colleague at work, Jessie Shefrin, notice how easily I had soldiered through what could have been a career catastrophe. Jessie asked me how I could get so easily detached from the real world around me—which as an artist she found impossible to do—and wondered if something was wrong with me. This moment had the effect of sticking a moth in my brain that jostled about until the bug finally broke me. Today I believe that spending too much time in the computational world is especially unhealthy for human relationships— because the people around you will start to seem trivial and literally living in a lower dimension. People are unresponsive when you command them to do boring things over and over, whereas in the computational universe there is absolute compliance and infinite scale (large or small). In many ways, the world of art is what has saved me, at different points in my life, from living in this heady world of cybernetic superpowers.

My first encounter with the reframing power of art was when learning foundational arts in my pen-and-ink illustration class. I had drawn a line a little too long. So my left hand moved instinctively with my thumb and pinky raised to valiantly chord the keyboard sequence for "undo"—only then to realize that there was no "undo" in the physical world. This phantom reflex is something I had to reprogram myself around. I also needed to recalibrate my relation to physical materials—something I only fully learned after countless cuts and calluses that make me feel lucky I still have all of my fingers. My favorite moment was discovering the hard way how much more difficult aluminum is to grind by hand than wood whilst willing an intricate shape manually over three months. Until then I'd dismissed aluminum as a wimpy "second-class" metal, but I learned to profoundly respect its coefficient of stiffness.

A similar series of aha moments came later in life, after art school, with my experiences in leading people. Having had the fortunate misfortune of experiencing more than a few intense failures in that realm, I've carried with me the accruing gift of knowing how little I really know about the noncomputational world. My ability to take apart and understand complicated machinery as an engineer has been a useful skill, but what I learned later in life about appreciating the enigma as an artist and the craft of collaborating with people as a leader has served as an important counterbalance. For that reason, when I encounter software developers who have made their path to becoming leaders, or artists, I'm especially excited to meet them—they are my family. Meanwhile, I've still snuck in a few hours of coding here and there to stay close to that world, because there's nothing else like working in its infinite spaces and feeling its magic—maybe as a form of therapy. If by chance I meet you someday and you see me acting like a machine, try to send a moth my way, please, and I'll hopefully break from my loop.

There's good and evil to be done with any technology. Consider how today we associate the name Nobel with peace, but we forget that Alfred Nobel invented dynamite, which has contributed to more wartime deaths than any other kind of weapon. On the other hand, dynamite also made it easier and safer for miners to clear tunnels. Or consider the Manhattan Project scientists who spent the rest of their lives weighing the balance between having invented nuclear weapons to destroy lives versus nuclear medicine to save lives. Today we can see a similar tension in the technology industry as it weighs the capability of educating anyone in the world with interactive videos versus manipulating people's lives by constantly tracking their whereabouts and influencing how they behave and act.

We can use our knowledge of computation to make complicated systems that sometimes have complex implications. Our brains can be trained to tackle the complicated pieces, but our values need to drive the questions around how we take on the complex aspects. If the work of developers and tech companies impacted only a few thousand people like back in the early days of computing, perhaps we could more easily leave those who speak machine to themselves in their nerdy basement studios or garages. But now that computing impacts virtually everyone at the ultrafine level of their daily micromovements and at the scale of the entire world, it is more urgent than ever to know how to speak both machine and to speak humanism.

5. Computers team up with each other way better than we do.

If you examine a personal computer from a few decades ago, you will notice that there's no place to plug in the internet. At first you might think, *Well, then it must have been using Wi-Fi or some other wireless technology back in the day.* Nope. Computers back in the day were islands mostly unconnected to each other. They were much less powerful, and much less useful. But over time we taught them how to connect to things like keyboards and printers and mice. And then we taught them how to connect to each other via networks that were "local"—near each other. The advent of the modem made it possible to connect via phone line to a computer that was much farther away. So a small, less powerful computer could talk to a bigger, more powerful computer, thus overcoming the limitations of its speed and memory. The small computer could become as powerful as the bigger one.

At MIT in the 1980s, I experienced this for the first time while sitting in the dorm room of an upperclassman friend of mine who was a computer whiz. He always made fun of my Mac computer as a kind of worthless toy, and so he wanted to show me what a "real man's computer" could do. So late one cold-pizza night I watched him as he deftly hopped from an MIT computer to a computer at Columbia University, and then over to one at Stanford. My response was one of wonder: how weird it was that we were in Cambridge, Massachusetts, but somehow also in New York and California at the same time. I soon learned the craft of jumping anywhere in the world where there were computers—but, mind you, back then there weren't that many machines out there to visit.

Hopping over to another computer started getting easier with AOL (America Online), but it didn't really get easy until the internet was fully democratized with the advent of the World Wide Web. The Web enabled computers to connect with each other and easily share files in ways that made sense to human beings. Access to every obscure thing, from a live image feed of a coffeepot at the University of Cambridge to a niche magazine called *Wired*, had soon become available on the Web. In 1994, there were 2,738 websites up by the middle of the year, jumping to 10,000 by the end of the year. In 2019, it's estimated that there are well over a billion websites, with that number continuing to grow. Say someone out there was seeking omniscience by learning about everything available on the internet—they would need extremely fast typing and clicking skills in order to visit all billion-plus sites in the world. Which is impossible, of course. But not impossible for—you got it—a computational system.

That's how Google is able to find what you're looking for on the internet—it simply crawls over all the websites in the world and compares what it finds out there with what you're looking for. It would of course be impossible for a human to do that, but it's as simple as writing code that looks something like:

```
for( website = 1; website <= all_the_websites; website
    = website+1 ){ < visit every page on the website and
    see if you get a match > }
```

This is an oversimplification of what actually happens, but you see how computation works at a scale and speed that make the pro-verbial needle in a haystack as simple to find as examining each individual strand of hay one strand at a time. What would be un-thinkable for a human being to do—like searching the entire in-ternet for the best "confused cat" article—is relatively easy to run as computer code. Making that search run quickly is the art and science of it all—and that's part of the reason why Google is worth so much money as a company—but the logic should make sense to you. So now is a good time to supersize our thinking by comparing a single person accessing the Web versus a single computer access-ing the Web on our behalf versus the sum total of computers in the world accessing each other's processing capabilities. That's right. If all the computers in the world are talking with each other at computational speeds and scales, then there's no comparison to be made.

Computers on the network are always cooperating with each other on our behalf, and on their own behalf as well, because there is value in having neighbors from whom you can borrow a proverbial cup of sugar. Similarly, computers are always talking with each other because when one doesn't know something, another one might. For example, when you ask your browser to bring you to the website howtospeakmachine.com, if it doesn't know where that server exists, it will ask another computer; and if that computer doesn't know either, it will just pass the buck on to the rest of the computers out there until a match is found. These communications and exchanges happen at speeds that far surpass our ability to type messages to each other human to human—they happen without delay and without resistance. Computers are always looking to collaborate in a way that is the ideal example of teamwork.

Smaller computers ask bigger computers to do favors for them all the time, like when you try to do something on your mobile device or from a home assistant appliance. If the computer doesn't have enough horsepower, it can simply kick it off to the network of computers living out there that we now refer to nebulously as "the cloud." It's important to remember that there's not just one cloud out there—most of the major technology companies like Google, Amazon, Apple, Alibaba, and Microsoft have clouds that consist of hundreds of thousands to millions of computers that are fully interconnected. These clouds require so much electrical power that they are often located near hydroelectric dams or solar farms in nondescript, windowless buildings with row upon row of computers placed densely in racks. With the cloud, the computer you have access to isn't just doubled in capability à la Moore—it's multiplicatively increased with each computer that's been added to the network.

Consider how networks of computers made of many machines—billions of them, if you add in smartphones—can all talk and connect with each other. They can all run loops singly, doubly, and in multiple dimensions, and they can text each other (and ask each other for help) millions of times per second. We like our computers to grow infinitely in their skill level and capabilities, and the cloud lets us do just that. Today we're at a point when holding any digital device is like grasping the tiny tentacle of an infinitely large cyber-machine floating in the cloud that can do unnaturally powerful things. The useless computer that sat innocently on your desk for many years as an exotic replacement for your typewriter is now a gateway to billions of other computers in the cloud. Once this fact settles in, the power of the cloud services companies will start to excite you for the possibilities they can bring and frighten you for the scale they operate at.

For example, the primary engine of computational power for Netflix, a video streaming service, is Amazon's cloud, because it is prohibitively expensive for Netflix to build its own and it is cheaper to rent the power in a competitive market. It doesn't make sense for a company like Netflix to own its own massive computing infrastructure when that is not a kind of expertise they currently need to own and control. Putting the issue of costs and expertise aside, the core advantage Netflix gains is the fluid scalability of computing resources and the ability to dial up or dial down their need to flexibly meet their consumers' demands. But when you note that Amazon controls half the cloud market and can easily change its prices at leisure, it only makes sense that Netflix would also want to shop the competitor Google's cloud for insurance. It also means that every cloud company in the world is working overtime to make their individual computing servers dramatically better at collaborating with each other to garner greater speed and power for their customers.

It's a relatively new thing that an entire tech company doesn't need to be built on top of its own computing power but can instead wholly rent it in a fully flexible, on-demand capacity. The cloud model represents a fundamental shift in how companies can get built, where the raw materials are all completely ethereal, virtual, and invisible. But that doesn't mean it's not comprehensible—it's just complicated. It's knowable and learnable. And in the back of our minds we need to be wondering what the future implications might be for servicing an entire race of machines to become better collaborators with each other than we ourselves could ever be. I know that my own uneasiness on this matter has made me resolved to devote the remainder of my life to foster teamwork and to partner with fellow human colleagues, because our computational brethren are beating us, exponentially.

Chapter 3

MACHINES ARE
LIVING

1 // Discerning what's alive versus what
isn't used to be simpler.

2 // The science of lifelikeness is
undergoing a renaissance.

3 // An artist's perspective keeps you
humanly curious.

4 // Life is defined by how we live in
relation to each other.

5 // Computers won't replace us if we
remain audacious.

1. Discerning what's alive versus what isn't used to be simpler.

On the first day of my seventh-grade biology class, my teacher Mrs. Figueroa explained how biology was the study of life, and you knew something was alive when you could see it "react to stimuli."

Seeing something move is the first clue that something is living. Think of how mesmerizing it can be to look at a lit candle, or even your image in a mirror—at first, you can't help but think what you're seeing is alive and distinctly separate from yourself. We react to shadows moving in a dark forest, we react to seeing faint movements in a still pond, we notice when a book falls from a shelf in a quiet room—we either connect it with something living, or something reacting to nature, or something supernatural like a ghost. All of these phenomena have some connection to the living world: animals, nature, the undead.

Scientist Valentino Braitenberg demonstrated how the mind can interpret lifelike behaviors from a world of simple robots composed of electronic building blocks. In this world, there are motors and sensors, and by plugging them together into different configurations, certain biological behaviors emerge from certain combinations. For instance, Braitenberg imagined a simple wheeled robot with a single motor to propel itself forward and a sensor to detect light. The vehicle was programmed such that when light was shining on it, it would move. Otherwise, it would stop. The more light shined on it, the faster it would go. The less light shined on it, the slower it would go.

Now imagine the logic in reverse. More light equals slower; less light equals faster. It's a vehicle that comes to a complete halt in the brightest light, but otherwise scurries forward when it's pitch-dark. If this vehicle was the size of your fist, you might see it as nothing extraordinary. However, if you saw the same vehicle at the scale of a penny, you would undoubtedly yell, "It's a cockroach!" What other creature loves to hide in the darkness and grime, and does its best not to move in daylight? Braitenberg went on to design other variations of lifelike behavior using his tool kit of parts to demonstrate even more sophisticated traits of living things, like aggression, love, and foresight. His vision lives on in part with the little vacuuming robots that skitter about on the floor by using simple behaviors that embody a kind of intelligence.

As for whether today we can be completely fooled by a robot and think it might truly be alive, even after close inspection—well, we're not there yet. But consider what the late philosopher Lewis Mumford said about the technology of electricity, the early twentieth-century equivalent of the internet:

> However far modern science and techniques have fallen short of their inherent possibilities, they have taught mankind at least one lesson: nothing is impossible.

So for now we don't have completely lifelike robots, but artificial intelligence is one space in which we're making unusual progress, thanks to a variety of advancements that we'll touch upon in the next section.

Consider a technology that has been around since well before the emergence of smartphones and self-driving cars: the automated customer service representative. It's the disembodied robot from the 1980s that everyone knows all too well. It's the synthetic life form on the other end of a phone line telling you, "Press one if you want to leave a message. Press two if you want to hear the directory." And much as a living human being would react, if it hears a one being pressed, then the computational robot will react to the stimulus, take your selection, and move it to an action. As anyone who has lost their patience when trying to navigate a maze of options with their numeric keypad will know, it's easy to forget that the automated telephone line will never get tired of routing you to the wrong thing. If you think you can get its goat by staying on the line for hours, you just need to remember the infinite loop that's running on the other end. The voice robot is indifferent to even the harshest of your indignant tones because it has no feelings. It literally has all the time in the world to repeat itself over and over again.

Interactive technologies of the past were simpleminded, missing many needed pieces, and could only respond slowly to our input. Because we roughly associate the speed of an object's responsiveness with "aliveness," a cockroach robot that takes more than a second to freeze its movement will be dismissed as not the real thing. A voice on the other end of a phone call that takes a few seconds to respond to your simplest request comes across as not the real thing either. In the old days, we could easily feel superior to the technology we were using because it was so slooooow—like when you would type something into an early search engine in the 1990s, and you would need to wait a while. But nowadays a search happens so quickly because it predicts what you might want to be searching for. And when you now talk to a computer, it responds as quickly as any other human being out there, complete with "umms" and "uhhhhs" so it sounds almost . . . human.

In the old days, it was easy to think that the computer was dumb because it took too long for it to do anything. These days we think it might be smarter than we are. That's because it responds so quickly to the stimuli we give it that it seems not only alive but also really smart. So, thinking back to what Mrs. Figueroa taught us about life, we expect something alive to respond to our input. Taking her lesson one step further, the faster it responds, the smarter we'll likely think it to be because it's so darn quick. And if it's got more stamina than we do, we're going to start feeling a little worried and insecure, because, for instinctual survival reasons, we don't like to be outpaced—especially by another species that we created to serve us.

When you consider the power of loops that prevent computational machines from ever getting tired while accessing an infinite cloud of capabilities, there's only one word that can describe these nonliving, pseudo life forms: zombies! And these invisible zombies should concern you for two main reasons: 1) you'll never win an argument with one of them, and 2) it will get harder to tell whether you're communicating with a zombie or not. The former will matter to you if you prefer to live your life with greater sanity and peace; the latter will matter to you if you wish to spend your life in an equitable relationship with what surrounds you. There are worse things than being made a fool by the zombies that are already multiplying around us today, but in the end I suspect you'll prefer to be among those capable of telling the difference between humans and robots, no matter how lifelike the latter becomes— or at least to have a fighting chance in distinguishing them.

2. The science of lifelikeness is undergoing a renaissance.

The 1970s harbinger of the potentially intoxicating power of computation, Dr. Joseph Weizenbaum, happened to be my professor in the AI course I took at MIT in the 1980s. I recall whispers from graduate student TAs that he was extraordinarily famous, but like any normal teenager I was more concerned with staying awake in class. Years later it hit me that he'd been the inventor, in the 1960s, of the famous Eliza program—the first computer program to simulate conversation in English. It was so convincing that it succeeded in fooling Weizenbaum's students into believing they were talking with a real human being.

We coded up a version of it the year prior in our beginning computer science class—and it was certainly a spooky, yet cool, feeling. The core of Eliza is a simplified simulation of what's called "person-centered therapy," which encourages the patient to find their own solutions using the original AI methodologies of the day. This involves "symbolic computation," where a set of words and numbers are arranged in a series of IF-THEN statements that execute in sequence. Symbolic computation is a fancy way of saying "writing computer codes that use more letters than numbers," which makes for programs that are less about calculations and more about processing higher-level information. The Eliza program "listens" to the words a person is typing, and then does one of two things: 1) repeats the words back as a question, or 2) triggers on certain words and generates a templated answer. Let's think of that as code being { hugged } within an infinite loop:

```
forever {
    < Take apart the sentence entered by replacing any
      instance of "I am" with "you are," and then add
      the words, "Did you say . . ." followed by a
      question mark. >
}
```

What happens? If you tell the computer, "I am having a bad day," it will respond, "Did you say you are having a bad day?" Or if you say, "I am going to go to the mall," it will respond, "Did you say you are going to go to the mall?" Pretty cool, huh?

We then learned to trigger based upon certain words. Consider a modification on the code hugged within the infinite loop:

```
forever {
    < Take apart the sentence entered by replacing any
      instance of "I am" with "you are," and then add
      the words, "Did you say . . ." followed by a
      question mark. >
    < Whenever someone mentions a relative like "mother"
      or "father" then ignore everything else and just
      say back, "Tell me more about your X." >
}
```

So if you tell the computer, "This is all my mother's fault," it will respond back, "Tell me more about your mother." Or if you say, "I've had it with my father!" it will respond back, "Tell me more about your father." If you throw in a comforting informal interjection like "Eh?" or "Ah!," the program can respond eerily more empathetically with, "Eh? What did you say? Your foot is bothering you? Ah! Tell me more about your foot."

Another simple modification to the logic in the code simulates actually listening to you and makes you think you're engaged in a deeper conversation:

```
forever {
    < Take apart the sentence entered by replacing any
      instance of "I am" with "you are," and then add
      the words, "Did you say . . ." followed by a
      question mark. >
    < Whenever someone mentions a relative like "mother"
      or "father" then ignore everything else and just
      say back, "Tell me more about your X." >
    < After the person's been typing for a while, refer
      back to something they previously said and
      extract the main subject, and then add the words,
      "I'd really like to hear again about X." >
}
```

If we play that out in an actual interaction, it will look like this:

```
You: I am having a great day.
It: Did you say you are having a great day?
You: Yes. It's been fantastic!
It: Tell me more.
You: It started with something that my grandmother
     said.
It: What does your grandmother do?
```

```
You: She has a bakery in Queens.
It: I'd really like to hear again about your great day.
You: Glad you asked! Let me start by . . .
```

Considering the sheer joy of conversing with someone who can tolerate your blah-blah-blah and give you the comfort of attentive listening, these kinds of conversations are validating and make us feel good. Furthermore, since this program is powered by an infinite loop, it will sit with you and listen to you endlessly, much like a patient parent or other loved one. When someone takes the time to listen to you deeply, you want to love them back. We love when we are listened to because it signals respect and acknowledges our existence—even when respect and acknowledgment is delivered by a machine.

The idea that a computer could so easily fool a fellow human being profoundly bothered the young Dr. Weizenbaum. While today we expect any ambitious technologist to immediately rush their invention to market without hesitation, Weizenbaum wasn't just any technologist. As a child, he had fled Nazi Germany and managed to escape the Holocaust. His worldview was grounded in an understanding of the dangers of misusing power. As a result, Weizenbaum spent his years after inventing Eliza speaking out about the dangers that computation posed to society, rather than profiting from or growing his career off of his groundbreaking invention. He presciently intuited that one day the computer would not only skillfully copy human mannerisms, but also successfully impersonate a human in conversation simply by knowing everything about the person it was talking with.

But there were three barriers to Weizenbaum's predictions:

1) there needed to be a way to collect all the information about an individual for the AI to respond convincingly, 2) there needed to be a way to collect many conversations with other people for the AI to learn new patterns, and 3) there needed to be a new way to process all the information gathered that symbolic computation could not yet achieve. The first two conditions have been met today by the advent of the smartphone, which has made it easy for the cloud to constantly observe all of our behaviors at an intense level, due especially to the addictive nature of our devices. The third condition has been recently answered by "artificial neural networks," a computing technique that was born around the same time as symbolic computation. Neural networks lost out in an epic research funding battle of the 1960s, but they now power the unusually accurate utterances of lifelike speech and writing that we're increasingly becoming used to from computers.

This last barrier was broken just a few years ago by a special kind of hardware: graphics processing unit, or GPU, which was designed to make computer games run faster and with more realistic imagery on fanatical gamers' PCs. Think of all the little polygons that get shaded for each pixel on-screen—this is a tedious set of operations that the main CPU could handle on its own, but over time it was reasoned that a special kind of "coprocessor" could be helpful to handle everything that appears on-screen. If you look at the specs for any computer today, the GPU will often be mentioned as a selling point, but it will only matter to you if you're a gamer. The GPU doesn't help to accelerate the standard data processing tasks needed to run your computer. It's only good for cranking out a massive amount of numerical calculations to blast cool graphics to your screen. And although we all love butter-smooth on-screen visuals, the utility of increasingly powerful GPUs for conventional computer users eventually outstripped their value and demand. But the GPU turned out to be the perfect inexpensive accelerator for the numerically intensive processing that would be needed to power that old AI idea of the "neural network"—dismissively left on the cutting room floor of AI discourse for fifty years and reawakened in the twenty-first century.

The term "neural network" implies two things if we look at the two words separately. Neural: it's related to the neurons in a brain. Network: it's about the interconnections between the neurons. A neural network differs from symbolic computing because it doesn't involve a series of logical steps defined by symbolic statements of code—it's more of a numerically "raw" relationship between an input and an output as mathematically modeled by a set of neuron-like elements connected to each other, like the network of neurons in a brain. Thus neural networks can't be easily unpacked as logical statements with symbols, like in a conventional computer program, but instead they involve raw numbers that pass across and between synthetic neurons over and over again until they "learn" a pattern.

An analogy is how your muscles will acclimate to a particular sport after you've played it for a while—you don't necessarily think of how you're doing it as a series of steps, but your muscles have indeed learned how to respond in ways that you've gained through hard-earned practice. In other words, neural networks are a means to encode the kind of "intuitions" we might have that can't just be written out as an easy-to-follow recipe. And it drastically differs from the long-accepted practice of writing lines of code composed of alphanumeric symbols. There's no actual computer code when it comes to a neural network—there's just a black box that learns patterns. Inside the black box is a rough mathematical model for how the neurons in a brain are thought to work electrically, which when stimulated the right way can learn patterns by making its own sparks, connections, and correlations with the raw numerical data that it gets fed.

So a confluence of factors sparked a resurgence of the neural network approach that had died in the 1960s but over time was resurrected with an unexpected "sonic boom" in computer intelligence: Moore's law made general CPUs faster and pushed the development of even faster special-purpose GPUs; the symbolic computing approach to AI had hit a ceiling in terms of how smart it could make computers; and all the data that's been gathered about us by the Googles, Apples, Facebooks, Amazons. When talking about this sea change, we tend not to use the term "AI," because it carries some negative connotations from the past. Instead, we prefer two terms to describe this newer kind of artificial intelligence: "machine learning" (ML) and "deep learning" (DL).

Specifically, deep learning is a technique used in machine learning. The traditional approach to creating AI was to teach a computer how to reason through IF-THEN rules. Deep learning, on the other hand, uses a model of the brain—neural networks in particular—to teach a computer how to think by observing a desired behavior and learning the skill through analyzing repeated behavioral patterns. For it to work well, the computer needs to observe our behavior. Preferably constantly and interminably. DL wasn't technically feasible due to the lack of succinctly large amounts of training data and the gargantuan processing power needed for the computer to become a worthy apprentice. Unsurprisingly, Moore's law has done the magic of bringing us enough computing power to calculate almost anything. We no longer need to explicitly teach it because it can just as easily teach itself with whatever data it has on hand, and it can reach out to the cloud when it needs more data to get even smarter.

Hold that thought for a moment—it's subtle and difficult to fully comprehend. It's a lot like when you visit a traditional French bakery, or *boulangerie*, where they offer two kinds of bread loaves that look identical. One is *pain au levain* and the other *pain à la levure*. For the average American like me, this makes no sense because they both have the same light brown, crispy crusts dusted with flour, as is characteristic of the French baguette. But to the French people, these two kinds of bread are completely different because of how they're made—which leads to subtle differences in how they taste. *Pain au levain* is made with a natural yeast, whereas *pain à la levure* is made with a chemical yeast—a simple mnemonic to remember the difference is that *levain* ends in "n" like in "natural."

I think of the traditional craft of making computing machines and old-style AIs by hand as akin to computation *au levain*—a nod to the natural handmade construction of codes. But the new methods of AI that are driven by neural networks are manufactured in a new, synthetic way akin to computation *à la levure*. Both types of bread, like both types of computation, satisfy the basic needs for which they are made—but they differ at the fundamental level of their respective ingredients. How do you tell the difference between *pain au levain* and *pain à la levure*? You can't tell by looking at the two loaves—but you can detect the difference with your nose: the natural yeast has a sour smell to it, whereas the chemical yeast has none. Likewise, because neural networks give off no "scent" while unnaturally doing things that were unimaginable even in the recent past, we need to pay special attention to this difference.

In Weizenbaum's era of AI, it was still possible to distinguish the robotlike responses of computational systems. It had a distinctive *au levain* smell to it. But even back in the day, in light of the progress of Moore's law, it seemed inevitable that it would become more difficult to differentiate between a real human's and a machine's response. If you are getting worried, like Weizenbaum did, about the implications of these emerging, uncannily lifelike neural network techniques, welcome to the ongoing debate about AI *à la levure*. Machine learning expert Andrew Ng describes the problem not as one of questioning whether a computer will eventually become awakened as a superior life form, but instead of: "If you're trying to understand AI's near-term impact, don't think 'sentience.' Instead think 'automation on steroids.'" Automation on steroids means we're living in an era where looping, infinitely large computing machines have provoked a kind of living material into action that excels at behaving like a zombie that never tires, AI is a mindless robot that does our bidding based upon patterns that its brain has been exposed to.

The new *à la levure* "living machines" work with far greater intelligence than they did in the past because Moore's law took us so far that unexpected things happened. GPUs led to a quantum leap in neural network processing power but Ng's "automation on steroids" required a large enough data set to feed the neural networks, otherwise they could never have become so proficient at learning on their own. For example, prior to 2012, the average error rate for image recognition was 28 percent, and for speech recognition it was 26 percent. After machine learning methods began to take hold, the average error rate for image recognition became 7 percent, and for speech recognition 4 percent. If the cloud continues to absorb more data from our activities, and if the zombies take no lunch breaks while they keep copying all of our moves, we'll eventually not be able to detect an AI versus a human. AI *à la levure* is here without the sour smell of AI of the past.

3. An artist's perspective keeps you humanly curious.

While I was the president of RISD, I saw it as my duty to champion arts education in an era that prioritizes STEM (science, technology, engineering, and mathematics) over everything else. Hearing how arts classes in public schools were being rationed from hours down to just minutes per week, and owing much of my technology career's success to the arts, drove me to wonder how to reveal the limits of a STEM-only approach. It forced me to consider the pros and cons of what Moore's law had brought forth, and it put me in touch with the economic benefits that occur when the arts overlap with STEM, convincing me that *STEAM* was a better strategy than STEM. Taking this theme to the US Congress in Washington, DC, and encountering countless stories of how art and design were taking a back seat to STEM was what subsequently motivated me to head to Silicon Valley. I wanted to understand how companies like Apple and Airbnb managed to fully leverage the combined capabilities of STEAM in their products and services. What did I find? That design-infused companies fully understood how art is the science of enjoying life, and thus, in order for their customers to enjoyably live with their products, they needed to involve artists in how their products were made.

Artists tend to get a bad rap—I certainly know that from personal experience, with my parents wanting me to study mathematics instead of art because they feared I'd never get a job. That's because art can seem esoteric and irrelevant, and yet it is where I have found a point of view that is fully compatible with the world of computation. Because the arts are not about what you can just see or sense; they're about discovering what underlies it all—about understanding what lies at the essential core of anything and everything. For example, artists know an important fact about apples that isn't immediately obvious when you bite into one.

Try drawing an apple from memory and you're likely to draw a circle with a stick on the top. But an apple is not based upon a circle or a sphere—it's more of a pentagonal solid. Go ahead and cut an apple in half, not sideways but in cross section. What do you see?

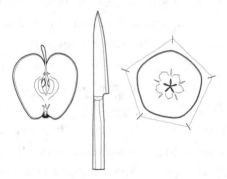

Do you see the pentagon? Knowing this fact will let you draw an apple more realistically. It turns out that the majority of plants on our planet have this fivefold arrangement—in the apple's case, it emerges from an apple blossom. The easiest way to verify this for yourself is to type the emoji for cherry blossom, which reveals an analogous fivetold symmetry. So a good artist knows this fact about the apple and draws from its natural underlying geometry— that you can't see unless you are really understanding (not seeing) intently. This is why artists learn to draw the human figure by first knowing the skeletal structure and how muscles wrap around bones as the first step to accurately expressing the human body. It's a logical and rigorous process.

We often mistakenly think that artists just depict the world the way they freely imagine it to be. I was certainly guilty of that mis- conception earlier in life, but the experience of being immersed in the arts allowed me to know the mind of the artist at a much deeper level. The artist's ability to hold both the foreground and background together simultaneously in their primary plane of thought is a skill I came to fully appreciate. I learned this by walk- ing around the RISD campus late at night while frequenting the Edna Lawrence Nature Lab—a menagerie of taxidermied animals, dried plants, and mineral samples as a miniature natural history museum of sorts. I would often go to stare and marvel at the mag- nificent cabinet of butterfly samples with blue wings, green wings, orange wings, all shapes and all shades on display in little glass cases.

One cold winter evening I stepped into the nature lab and found it empty, so I felt comfortable asking the student attendant if the butterflies were the most popular items there. They were my fa- vorites. To which she quickly replied, "Nope. The gourds are." And her quizzical expression said to me, *You like the butterflies? That's*

SO obvious and childish. There's a good reason why we're all intimidated by college students who are smarter than us, and so I sunk a little into a servile slouch to tacitly express, *Please teach me, O master!*

Luckily, she threw me a bone and quipped, "But the feathers are a-MA-zing!" At which point I became slightly worried about what she might have been smoking on her break. *Feathers?!?!* I thought to myself.

Noticing my abject confusion, she grabbed a butterfly sample out of the grand display case and whisked me off to one of the high-power microscopes we had just installed. "Here. You see!" I looked at the imaging display come into focus and saw none other than . . . featherlike scales on the surface of the butterfly wing. It's the powdery substance that gets on your fingers when a butterfly unfortunately gets caught by your younger inquisitive self and sadly can no longer can fly. But it's not actually powder; instead, up close, it looks like feathers. I slipped back into the cold winter air, but warmed with edification from this inspiring student's ability to connect across vastly different scales so fluidly. She reminded me how artists excel at making unlikely connections all the time.

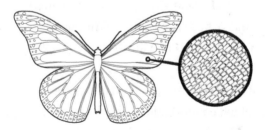

My practical respect for unlikely connections has compelled me to curate my own little desktop nature lab with a display of

Conus textile shells. These sea snail shells aren't feathered by any means, but sport a spiderweb-like pattern of lines that form little triangles that fan out into more triangles that continue to multiply—much like when you might zoom into a Koch snowflake to reveal a never-ending repetitive pattern. This pattern resembles a computer algorithm discovered by renowned computer scientist Stephen Wolfram called "Rule 30" and draws a surprising link between our organic physical reality and the invisible world of computation. So the seemingly inhuman algorithms that we code on computers may be more human and natural than we'd like to think.

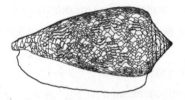

Rule 30 works by taking a long row of numbers set to zero:

00000000000000000000000000000000 . . .

And then writing a simple piece of code that transforms the next row of numbers based upon the following eight rules:

```
Rule 1: If there are three 0s in a row then make the
        central cell 0.
Rule 2: If there are three 1s in a row then make the
        central cell 0.
Rule 3: If there is the sequence "110" then make the
        central cell 0.
```

Rule 4: If there is the sequence "101" then make the
    central cell 0.

Rule 5: If there is the sequence "100" then make the
    central cell 1.

Rule 6: If there is the sequence "011" then make the
    central cell 1.

Rule 7: If there is the sequence "010" then make the
    central cell 1.

Rule 8: If there is the sequence "001" then make the
    central cell 1.

The outcome is pretty boring for our row of numbers when set
to zero because if we run this algorithm for a few cycles we'll get:

00000000000000000000000000 . . .
00000000000000000000000000 . . .
00000000000000000000000000 . . .

But if we set one of the numbers to 1 in our starting row, something starts to happen with each step of the algorithm:

00000000000001000000000000 . . .
00000000000011100000000000 . . .
00000000000110010000000000 . . .
00000000001101111000000000 . . .

Notice that a pattern of ones and zeroes is growing, such that if
you were to loop over a few hundred cycles you would end up with
a pattern streaked with peaks and valleys. Replacing the number 0

with a white square and the number 1 with a black square results
in a pattern that looks like this:

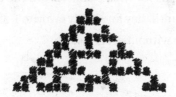

And by the tenth row my hand has grown tired, so I write a program to tack on 290 more rows with perfect rectangles instead of
my hand-drawn squiggles—it did the job in less than a second.

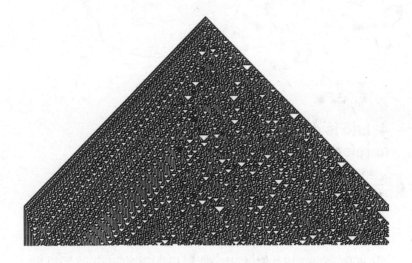

The fact that the pattern shares a striking resemblance to the
Conus textile has been the subject of multiple scientific papers that
ponder what it means—and if you're superstitious like me, you'll
wonder if it's more than just a coincidence. Does nature speak machine? Or can machines speak nature? Are we just machines?

You're certainly thinking like an artist if you are asking these questions too, because artists know to look deeper than just surface beauty. They dig for what's underneath the underneath, and they will not rest until they find it. My favorite T-shirt graphic says it best: "The earth without art is just 'eh.'"

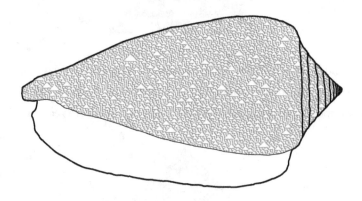

4. Life is defined by how we live in relation to each other.

I work with a lot of creative people who look down upon "noncreative" managers and businesspeople because they don't see them producing anything with their own hands. There's a bias for artists or designers to see themselves as making something with integrity, getting their hands dirty, and bearing the evidence of having done honest work. These dirty hands are in stark contrast to the cleaner hands of folks who spend most of their time talking rather than making.

This "talker versus maker" dichotomy was introduced to me by a design student I first met during my regular walks around campus at RISD. With characteristic youthful swagger, he sidled up to me on the street and said, "Hey, John. I noticed you were talking to those students over there just now. You shouldn't talk to them." I was a bit puzzled, as I believed in inclusivity, and so I asked him why. He replied, "Because they're talkers. They talk about their work and don't spend time making the actual work. You should hang out with us instead. The real makers. The ones who are always making instead of wasting our time talking."

The student's frank and simple logic reminded me of a similar mentality shared by engineers at MIT in how they saw their business colleagues at the MIT Sloan School of Management. People who talked were considered "evil" and those who made things were considered "good." But as a maker who had fully embraced his inner talker, I couldn't help but tell the young man, "I'm sorry to tell you this, but you're a talker yourself. Otherwise you wouldn't have walked up and talked with me. Let me tell you why it's good to talk. It lets you connect with other people besides yourself—and in the process, you not only find the answers to what you don't know, you also increase the potential of what you are making because you've made a new friend." I was happy to see this student later become active in the TED community once he'd discovered his maker *and* talker sides. Yet my intent here isn't to focus on him but to consider the paradigm of *doing* work versus *connecting* work.

Designers (and developers) will identify themselves as makers when they're first starting out. They use their hands to get work done, and their pride comes from the recognition generated by the work itself, rather than by what they are forced to say about it in a public forum. Because we as makers tend to live inside the majesty of our "mind palaces," where our discourse with ourselves and the work we're deeply engaged in are the only things that matter. Or, as described more cynically by the late David Foster Wallace, as being left alone to "the freedom to be lords of our own tiny skull-sized kingdoms, alone at the centre of all creation." You might detect a tone of disdain in my description here, but it's not a sentiment of disrespect. It's my home turf, where I feel the most comfortable. I call it out this way in a regular ritual of forcing myself to become more balanced and to interact with the world. Because for most of my life, I loved making things. Period. And I didn't really have to do any talking about it. In fact, the epitome of this mind-set is one of my most well-known artistic works from my early career, a haiku I composed to that effect:

> All I want to be
> is someone who makes new things
> and thinks about them.

There was a time in my life when I could live as a pure maker. While my early computational design work was gaining notoriety in Japan, I had a publisher named Naomi Enami representing my work. Mr. Enami would do all the talking for me, and I would do the making behind the scenes.

One fall in the early nineties, there was an influential conference at which Mr. Enami wanted me to present. I was terribly shy and didn't want to speak in front of people. In the end I had to acquiesce, because Mr. Enami had helped me so much, and so I nervously did so. That evening, Mr. Enami suggested that I go to the after-party and talk to people. I told him that I didn't want to, that I preferred to go back to my room and continue writing my code on my laptop computer. He looked at me a bit sternly and said, in Japanese, "Mr. Maeda. Making relationships is just as important as making software or making designs." With youthful arrogance, I of course didn't listen to him, and I went back to my room and started coding. A few weeks later, Mr. Enami went into a coma. Soon after this episode, I realized that nobody would be talking for me anymore, and that I would need to learn how to talk for myself. So I made a conscious choice to step out of my "skull-sized kingdom" and become a talker while still endeavoring to make as often as I could along the way. In the process I've encouraged makers to learn how their work can benefit from collaborating with talkers, because talkers don't just talk with each other—they're talking with prospective customers too. The connecting work done by a talker is equally important to the making work itself, since it increases the odds that there will be an audience greater than just the makers themselves.

Whether in the real world or the computational world, connecting work is the catalyst for making changes happen at a scale that's larger than just the *doing* work that an individual can perform. The early computer by itself was hardly more than a deluxe calculator that annoyingly monopolized the screen of your television set to make it work, but later, with the advent of the internet, it became something amazing by becoming interconnected with other machines. Recall how there are billions of computers to choose from when we're trying to reach out to howtospeakmachine.com, and how we don't want to wait more than a second to access it—we can thank domain name system (DNS) technology for making that miracle happen, with many computers collaborating with each other. The site name is tossed over to a namespace server, and if it doesn't know the right mapping, it will then reach out to another one, and another one by looping forever until a match finally comes back—all within the span of a few hundredths of a second. Whether it involves machines or humans, there's magnificent power when depending upon each other, because we can together do things that we could never achieve alone. Or, as once expressed by the Shawnee chief Tecumseh: "A single twig breaks, but the bundle of twigs is strong."

In the mathematical world, the codependence of life was eloquently expressed as a succinct algorithm in the 1970s called "Conway's Game of Life." With "Game of Life" in its title, Conway's "Life" is immediately misleading, because it brings to mind the better-known family board game invented in 1960, which evokes a much more playful experience than the stark grid of black and white squares upon which Conway's game is played. Furthermore, this game isn't designed for you to have fun, because it's more of a mathematical game—which will probably only amuse you if you are mathematically inclined. Lucky for you, your journey into the computational world, which is largely based on mathematics, will help you appreciate this important gem. At least I hope so . . .

Conway's "Life" is applied on a grid similar to Wolfram's Rule 30, but in a two-dimensional grid of cells instead. There are just four mathematical rules, where each rule simulates an overly simplified model for how a life form might interact with neighboring life forms "living" on a grid.

Rule 1: If a living cell has only one living neighbor,
it subsequently dies from loneliness.

Rule 2: If a living cell has a few living neighbors,
then it's stable, and will subsequently continue to
live.

Rule 3: If a living cell has too many living neighbors,
then it subsequently dies from overcrowding.

Rule 4: If a cell is empty and it has a few living
neighbors, then mating will result in a new living
cell to be subsequently born.

These rules can be played out by hand and step by step with black and white paper chips on a table, where black means a cell is "living" and white means a cell is "dead." You can imagine it getting quite boring after the second or third iteration on a large grid. Fortunately, by the 1980s, computers had become powerful enough to sit in a loop while covering as large a grid as desired. I still remember when Professor Abelson came into one of our computer science classes back in the 1980s, absolutely giddy with excitement to show us something miraculous that had become visible for the first time. He showed the little black and white squares blinking on and off. And then slowly, over time, we could see a few group behaviors. Little groups started to move around together. Little groups started to blink together. Little groups collided into other groups, and new little groups sprang up. But for something as grandiose as the so-called Game of Life, I was expecting a lot more than little black and white squares blinking on the screen. It went completely over my head, and I dismissed it as too hard for me to grasp.

What I recognized many years later as the big idea was the simplicity of it: that just four rules applied to local relationships between a cell and its neighbors could somehow create group behaviors where the individual cells act as a pack. Because nowhere in Conway's rules is there the idea of a group of cells. So when a so-called glider pack of cells breaks off and starts to move around on its own as if they are a single entity, that's an unexpected outcome.

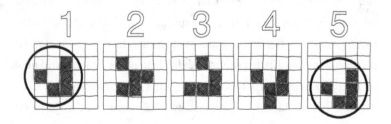

With just four rules pertaining to a single cell that communicates with other adjacent cells on a grid, multiple cells can behave together as a collective without any knowledge of the enclosing environment. It's as if communication with each other is what makes this kind of collaboration emerge.

When we work together, we find collective benefit by attaining abilities that no individual can have alone. For example, emperor penguins huddle next to each other for warmth in a clever way, constantly taking turns rotating their position at the warmest center and the chilly edges. They do this by skittering about on tippy-toes to decrease the surface area of contact with the cold ice, while being sure not to touch each other and thus avoid compressing their feather insulation layers. We can see a similar kind of emergent behavior in how birds and fish move in tandem with each other to create beautiful flocking configurations. Nature's intent isn't to create beauty, but to ensure greater survival with such herdlike behavior as the proverbial "bundle of twigs." And given how we now know that the human body comprises trillions of microorganisms that outnumber human cells by a factor of ten, we can think of our entire being as one big collaboration factory at the tiniest of scales.

There's a lot of talking (and making) happening in our bodies right now. It's happening in the computational world right now too. The fact that computers can talk with each other means that they are collectively as smart as the most brilliant computer that they can access—just as the social networks we humans occupy now connect us. Will we figure out how to work with each other as fast as computers have learned how to work together, talking with each other and leveraging the full power of our collective intelligence? Perhaps. But only if we can learn to listen to each other to start. If we can perfectly train computers to listen so well to us and so fully among themselves, shouldn't we be able to do the same for and with our fellow human beings?

5. Computers won't replace us if we remain audacious.

When we look back over fifty years ago to the birth of AI, the original goal was to make a computer think—where masterful "thinking" meant playing the game of chess, because that's what we nerds do. The computer science field was desperate to make itself seem more legitimate, and chess is associated with a certain kind of logical mindedness. So putting chess in the spotlight of computing research was a strong recruiting game for garnering the interest of smart mathy folks. But the promise of AI as powered by the chess players' rhetoric was especially challenging when trying to convince four-star generals to back and fund their new field. Ultimately, their marketing efforts prevailed and AI research prospered for almost two decades starting in the 1950s. But the high expectations set for achieving intelligence with the primitive computers of that era eventually resulted in major funding cuts to research in the 1970s, known as the "AI winter." Fast forward to today and we know that summer has come back to AI—because of the new AI *à la levure*—with the return of government and industrial funding at the scale of billions of dollars.

While looking back through the history of computing in writing this book, I couldn't help but to find myself arriving at MIT in 1984, the same year that the Apple Macintosh was born. That year coincided with the very moment when the billion-dollar AI industry was starting to collapse in the wake of the AI winter—made evident in the accompanying shortfall in research funding. It helped explain to me why, when I worked in the MIT Artificial Intelligence Laboratory as an undergraduate, my fellow classmates all tended to look at me like I had picked the wrong horse to bet on. For betting on AI back then was much like betting on penny slot machines, since computing power was still insufficient to show any meaningful progress. The grand task to make a computer beat a human at chess seemed far less important than enjoying the superior work productivity of an AI-less word processor that could replace the clunky typewriter sitting on our desks.

Yet fast-forward to today, and somehow AI is a topic in every Silicon Valley conversation—where an intelligent machine is considered an eventuality—not to mention in business circles across all industries and in the political sphere as well. The speed at which AI is developing reminds me of what the financial community calls the "Warren Buffett rule" of compounding interest as the key factor to his investing success. Behind the rule of compound interest is the simple idea that the more you save, the more you get to earn over time. So even at a low interest rate, where you'd expect to see corresponding small returns, if you keep adding those returns back into your savings, the interest you earn will compound over time. So the few pennies you might earn on interest in the bank in one year may not matter, but it's the compounded interest after a few decades that turns those pennies into dollars. For example, if you held on to a penny for fifty years at an annual interest rate of 1 percent, you would end up with a total of 2 cents. At a rate of 5 percent it would be 11 cents. And at a rate of 10 percent it becomes $1.17. But if you use the Bank of Moore's Law, conservative annual interest rate of 66.66 percent, that penny you held over fifty years would accrue in value to more than a billion dollars, at $1,234,721,113.27.

By now, as a visitor to the computational realm, you're at least a cautious believer in the concept of exponential growth. But computer folks have always known about this unusual, invisible phenomenon of compounding gains when working and thinking computationally. For them, science fiction is not just fiction, but instead the most logical means to paint plausible realities. In the early 1990s, science fiction writer and computer scientist Vernor Vinge wrote a paper in which, in the first line of the abstract, he predicted, "Within thirty years, we will have the technological means to create superhuman intelligence. Shortly after the human era will be ended."

For non-computer folks, it would be easy to dismiss this statement, but for a Moorean thinker it would feel like a completely logical outcome. Vinge coined the term "the Singularity" to describe the moment of "imminent creation by technology of entities with greater than human intelligence."

A few years after Vinge's paper was published, renowned inventor and scientist Ray Kurzweil went one step further by writing a 672-page book on the subject of the Singularity, describing exactly how it might happen. Kurzweil predicted that by 2015 computing power would surpass the brainpower of a mouse, and that by 2023 it would surpass that of a human. He went even further and predicted that by 2045 there would be more computing power than all of the human minds combined on earth.

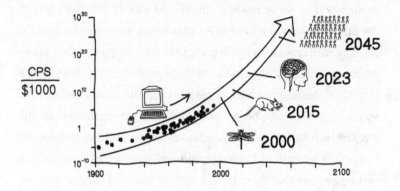

Suffice it to say, this Singularity stuff will come across as science fiction to anyone far removed from the world of computing. But for AI pioneers like the late MIT professor Marvin Minsky, who preferred science fiction over general literature for its ambition to imagine big ideas, and for others like him, the Singularity is just another big idea that is destined to happen. The exponential progress in computing technology over the last few decades indicates that the Singularity is not a far-fetched science fiction future. There is even a Singularity University in Silicon Valley, co-founded by Kurzweil, to investigate the impending future when machines will eventually surpass us. Given that in 2015 a mouse brain was reported to have been simulated on a supercomputer, as Kurzweil predicted would happen, we can wonder what might come of Kurzweil's prediction for 2023.

Already, you should be able to sense that something is different about the computing experiences around you compared with the previous decade. It's the difference between computation *au levain* and computation *à la levure*, but you've likely failed to notice the quantum leap we've made from AI that smells like the computers of yesteryear to a kind of AI that bears no computer scent at all. The millions of invisible computational loops running somewhere out there in the cloud—without ever slowing down or getting tired despite running at close to infinite scales and infinite levels of detail—have been compounding in strength and capability. The technology industry and investing world have endeavored to fuel the growth of these magnificently powerful systems that no longer resemble the cold, calculating material they're made of, as they now draw upon the observations of artists and their deep understanding of humans and their environments. As a result, we'll likely experience more computational surrogates that blink, smile, sashay, express uncertainly with a few "umms," and even flirt with you to get your attention. They'll be eminently likable because they'll never give up trying to improve, and they'll be able to draw upon increasing knowledge about you every second of the day.

We know that computational machines excel at copying our behaviors, they excel at collaborating with each other, and they're good at both large-scale and precise details. The immediate thought you should have in mind if you are a boss is, *You're hired!* Closely followed by, *But will you be taking my job eventually?* The way to deal with the latter concern is to remember that the goal of a good boss is to become expendable so that your team will grow faster than you. So, in a sense, the answer that should give you comfort as a good boss is *yes*. Next, remember that you'll never be completely expendable if you're always outgrowing your own capabilities—this is what makes us difficult to copy, because we're moving targets. So do your best to be audacious, like the younger generation you work with, and be unafraid to leap into the unknown with their same audacity. Then, as you recognize the magnitude of the challenge before you, draw upon your courage and experience to do as my late mentor William J. Mitchell would often say: "Blast away at it. Get it done!" A warrior can't be a worrier. Run toward the challenge instead of running away from it.

I have been extremely lucky to be deeply immersed in the world of speaking machine for most of my life. Having had the opportunity to think through the next set of ideas I'll be sharing with you in the three chapters ahead, I feel an awakening inside me. Today I feel an imperative in our need to rethink the implications of computation in the design of new products and services, because we're at a turning point that will irreversibly impact the future of humankind. We are currently on course to reach the Singularity in a completely inequitable way for the many human beings who don't speak machine. At our current rate of progress, and with a minority of computing haves who will lord their vision over the computing have-nots, we will permanently chisel into the cloud the biases of a narrow band of fluent creators.

It feels like now is the last chance we're going to get before *plop* and the entire pond is 100 percent covered with lily pads. So let's turn toward the implications for products created with computational machinery that use loops, grow large, and are increasingly more alive. Companies like Google, Apple, Facebook, Amazon, Microsoft, Alibaba, and others have all long been in this space and have had a huge head start. You're being courageous—no longer just audacious. Let's now learn how computational products are fundamentally different and are in desperate need of your new-comer perspective. We need to get you baking your own signature loaves *au levain* and *à la levure* to ensure the fullest kind of representation and diversity of thinking that the future needs. The Singularity hasn't happened yet, so let's get baking!

Chapter 4

MACHINES ARE

INCOMPLETE

1 // Timely design is more important than timeless design.

2 // The Temple of Design doesn't rule this century.

3 // Perfection isn't as good as understanding.

4 // An incomplete idea is only a good one if you iterate.

5 // Emotional value is a need-to-have instead of a nice-to-have.

1. Timely design is more important than timeless design.

Software developers use odd lingo. If you hang around them for long enough, you'll hear the word "agile" and wonder why more programmers aren't wearing athleisure clothing. You'll hear "lean" and ponder the candy bar wrappers everywhere. And "scrum"? Maybe some sci-fi movie you didn't see? These terms all essentially describe a computational philosophy dedicated to shipping an incomplete product followed by as many revisions as possible, instead of trying to ship a product that is complete.

This approach is relatively new. Traditional software product development is managed in what's called a "waterfall development approach" as a logical series of steps that starts from high above and then goes through a series of transformations along the way down to completion, like the flow of running water as it falls off a cliff. Waterfall is a linear process involving painstaking attention to the product's requirements, followed by design, then implementation, then testing, and then maintenance. This is the same way you'd build any normal physical product like a car. But it's not possible to make revisions to a car after it's been delivered, unless you issue a product recall, which is both cumbersome and damaging to a company's reputation. Sometimes, this waterfall can take years to flow to the ending point where the product finally gets built, and with comical results. For example, a designer in the car industry once told me how one of the big car companies predicted that all cars would have fax machines. As a result, the interior cabin was designed with a special compartment for one. A few years later, when the car was finally being built, fax machines hadn't become as ubiquitous as expected, resulting in a large empty space in the finished product that no longer had a purpose.

In the past and even in the present day, assembly lines are designed with the singular goal of reducing the additional cost of each unit, or marginal cost, in order to achieve greater economies of scale and increased profit per unit. Launching a digital product falls under the once mythical business categorization of "marginal cost close to zero"—which is an exceptional opportunity, because it means there's no downside to building and distributing hundreds or thousands of extra copies of a digital product. In addition, you never have to deal with carrying inventory, managing its movements, and all the associated costs. For those with a traditional manufacturing background, it is a considerable conceptual stretch to imagine that such a plum business model could ever exist—but it's easy to do so when considering what you now know about computation as a raw material that defies our laws of physics. So defying the existing laws of business should not only feel easy, but it should feel logical too.

This unique property of computational products means not only that their production and distribution costs are financially advantageous, but that product development costs can be significantly lowered by making a choice to never ship a finished product. You can always "replace" the product digitally with a brand-new and incrementally improved one, and remove all the financial risk from investing heavily all the way to a finished product. What's more, the ability to remotely monitor how people are using these products means their makers can easily nudge the product's design toward whatever will work best for the end user.

In a world that's made purely of computation, it's trivial to ship a change so fast that you often don't notice that it's happened. It should be no surprise that quickness is a fundamental aspect of quality in the computational age—"good" is defined by how fast a new feature can get to you and how quickly you can get it to work for you. So especially when we consider the doubling effect of Moore's law, we should expect that all digital products should be twice as fast as they were a little over a year ago and cost the same, if not even less. Consider how on any given day you are explicitly upgrading your software to "the next version" or waking up to a new version of the system that just last night might have been something quite different. The computational products you're most often using become cumulatively better each day, but you might never have noticed the improvements being made because the many incremental changes are compounded over time.

If you're looking for a metaphor to explain this shift best, just point your finger toward the sky. That's right. It's all about our constant connection with the cloud. Because the cloud has enabled a new kind of agility when it comes to making, shipping, and improving digital products. Because every networked device is advantaged by a constant connection with the cloud that allows data residing in anything anywhere to be deleted, refreshed, and synchronized. And because the cloud and every computing device attached to it are always changeable and ready to be refreshed, it's contradictory to put in the effort to pursue making a perfect "final" version of a product that is worthy of going into the permanent collection of MoMA. The cloud has made the pursuit of timeless design irrelevant—what matters instead is to be timely. What matters more is to be evolving in real time and never stop. And to seek bliss and satisfaction in a state of always being obsolete.

This might vaguely sound like a form of "planned obsolescence." This business strategy, made popular by GM after the Great Depression, tried to get consumers to upgrade their cars by making them feel like their current one had gone out of style and was no longer appealing. The strategy of preying on customer insecurities ran counter to the prevailing wisdom of Henry Ford's more pragmatic approach to offer one car in any color "so long as it's black" and with a commitment to providing long-lasting value. But GM's business-oriented approach decidedly won over Ford's honest-engineer approach. Over time, planned obsolescence became known by environmentalists as an evil of capitalism because it resulted in consumers throwing away products faster to replace them with newer, better ones. But what makes software different from the planned obsolescence of automobiles is the computational medium's intrinsic property of being "always obsolete" and just one update away, and with minimal environmental impact, to be transformed into the newer, better version. In fact, the standard Silicon Valley–style expectation is for software products to constantly get improved and upgraded while you're asleep or even while you're using them.

There's an insidious side to this strategy as it is played by Apple and other companies that we're just now catching on to—which is to purposefully make incremental improvements in their software that require hardware upgrades to function properly. Either you're missing out on a cool new feature or otherwise your device will run sluggishly because it needs more memory or a faster processor. For the longest time Apple was the only company that could play this game, but Google is catching up with its Pixel hardware platform. When a technology company owns both the software and hardware, you can expect to become hooked on the latest version and newest models way more efficiently than any car company could have ever dreamed—and you'll pay to keep up accordingly, like clockwork, with each financial quarter.

2. The Temple of Design doesn't rule this century.

If design is reasoned intention, then we need to change our reasoning skills. But here's the problem: the traditional world of design doesn't live in the cloud, but in your beautiful living room. It's the antique chair that your grandmother gave you. It's the Swedish flatware that you got as a wedding gift. It's the French leather case that rests atop the aluminum table you spent more than a few months mulling over. It's the fashion magazine you flip through while sitting on Grandma's chair telling you that your striped outfit needs to be thrown into the fire because artist Yayoi Kusama's polka dots are now all the rage. This isn't the world of the cloud at all, and yet it's what we love most about design. Museums and galleries are filled with these objects, experiences, and highly opinionated prompts that we treasure in a way that can best be described as worshipping the Temple of Design.

The Temple of Design isn't an actual place, but more of a standard for knowing what great design is. In harsh, concrete terms, it is a cartel run by the museums, galleries, fashion and beauty industry, and arts education institutions worldwide that decree what great design is all about. Which is . . . what *they* say is great. And if you can't tell it is great, then, well, you aren't a designer. Back when I was trying to figure out design, I fell prey to this conclusion that I wasn't a designer despite my freshly minted MIT degree, so I went in search of the Temple of Design by simply going to art school for a degree, and then much later actually running one of them.

I can save you some time by informing you that becoming a valid member of the Temple of Design has just two requirements: 1) you need to understand the history of art and design, and 2) you need to think and work like a designer. The former requirement can be met if you read a lot, visit many museums and libraries, and simply put in a few thousand hours of learning. It is painless and eminently pleasurable. What makes it even easier these days is that you can walk into a design museum with a smartphone—so you can learn a lot faster than it took me. I spent four years in the arts and design library at Tsukuba University to figure out things that, if I had had Google at my fingertips, I could have learned in a fraction of the time!

The other Temple of Design requirement—thinking and working like a designer—is significantly harder than googling your way through design's history. But knowing your ABCs of the Bauhaus school will at least get your foot in the door of the club, so it's a worthwhile foundational investment to start. But to actually become a designer, you'll need to actually design a few things with your own hands. You're going to make some horrible things, which you won't fully realize until a trained designer looks at what you've done. Learning the minimal basics of design requires only tens of hours of practice, but being able to excel and make your way as a real-deal designer requires the proverbial ten thousand hours of deliberate practice that anoint you as classically trained.

As someone who's made it through design boot camp, I can tell you that there are three kinds of design styles at play that are embedded in the problem that you choose to solve, or embedded in the manner by which you approach your solution:

1. Out of style: This means you are too late. The crowd has left the moment. You weren't paying attention and are now passed by. Expect to be slightly shunned at first, but if you remain strong and have conviction, you just might defeat the odds.

2. In style: This means you are right on time. You are comfortably ensconced in the majority. You had to earn your place in the pack by decoding which elements of style defined the moment. You belong, you conform, and you feel relieved for now. But beware that the next wave will soon come. So be on guard.

3. Making style: This means you are in the vanguard. A minority has stepped forward to stand athwart the majority and to make a statement that may contradict what's popular. There is no guarantee that those who make style will succeed in becoming "in style." When you have the guts to stand out and be smacked down, that's often the definition of truly cool.

The folks who get to make style are card-carrying members of the Temple of Design. It's a comfortable place with heroes and a well-determined hierarchy. It is a world of advantaged and wealthy people who determine what is desirable by virtue of simply having what others do not have because it's beyond their reach. It's a tradition that reaches back to the way royals distinguished themselves from commoners—by having what others could not have. So it isn't a coincidence that the US online retailer Design Within Reach is often jokingly referred to as "Design Without Reach." It is a world of unfair advantages constructed out of influencer opinions, and thanks to the pomp and circumstance machine of the Temple of Design, it can seem extremely important from the inside. Of course, I love being there for all the reasons you can imagine, but I also have a responsibility to point out that it has fallen out of touch with the computational era.

The Temple of Design has long been able to dictate what making style is all about, because making anything for manufacture has always been a capital-intensive task. The question of what would go to production rested on the knowledge, and network, controlled by the tastemaker hierarchy. But this reliance on the rulings of a privileged few has long been under siege by the data-gathering and informal networks of rationally minded, non–Temple of Design acolytes. And the financial and influence capital made available in Silicon Valley, combined with the computational power now available to anybody, means that rationality can prevail and quality design can be had by all for an exceedingly low price—as long as it's delivered on an electronic device. We live in an age when you never have to be "out of style" and can always be "in style" because you can always be equipped with the latest and greatest computational product designs—which you can now, in turn, play a key role in helping to determine. And nothing the Temple of Design can say or do really impacts what happens with computational products today, because, frankly, the Temple of Tech has sped past it in relevance at Moorean speed.

The doctrine of the Bauhaus school was developed in response to the industrial revolution and the newfound ability to manufacture products using machinery and assembly lines in factories. Affordable appliances could finally be produced for the masses, but they were often difficult to use and didn't fit the décor of consumers' living spaces. So in the early 1900s the Bauhaus education program introduced superior ways to make products that were usable, desirable, and affordable—but the resulting work was not immediately met with popular acceptance. For that reason, the words of the Austrian writer and Dadaist Raoul Hausmann were especially relevant for that early period of rapid transformation:

The new human should have the courage to be new.

It follows that, a century later, the old Bauhaus ways must be shed to usher in the new ways afforded by the new industrial revolution now under way, powered by computation. The new courage required is to steel oneself to take on the challenge of speaking machine, as you are doing right now. Consider yourself a member of the new Bauhaus of this century.

Micro-improvements for computational products are constantly refreshed on a schedule as rapid as by-the-second, by-the-day, or by-the-month, where the metric of quality is far more about how often it changes than about maintaining an unchanging product that pretends to be timelessly perfect. This runs counter to the ways of the Temple of Design and its tradition of laboring over decades to consciously evolve what should end up in the history of made objects as "finished" pieces to be worshipped forever. Instead, today we live in a real-time reality grounded in computational power fed by the clouds that have cast a shadow over the golden facades of the Temple of Design. We need to fundamentally renounce the traditional notion that design ought to aspire to completeness.

So the new definition of quality is the opposite of the Temple of Design's definition of quality: a finished product painstakingly crafted with integrity. The new definition of quality, according to the Temple of Tech, is an unfinished product flung out into the world and later modified by observing how it survives in the wild. It's the opposite of betting big on a brand-name designer vetted by the Temple of Design to get it heroically just right with a grand ceremonial debut. And instead it's about gathering a team of talented nerds who can take full advantage of an alienlike material with new properties that can transform business. It requires a seismic shift toward an attitude of being "lean" (eliminating waste and favoring experimentation over perfection) and "agile" (flexibly responding to customers' changing requirements). Or, to put it in less techie terms, quality is about proudly embracing the attitude of working incrementally and completely underwhelmingly—to send oneself on a never-ending journey of making products that can never make it into the collection of the Victoria and Albert Museum. So just listen to Hausmann's sage words and let's all have the courage to be new!

3. Perfection isn't as good as understanding.

For a computational product designer, waking up each morning and asking yourself how you can lower your high standards is an odd way to get started. You must accept that you're not going to be drawing upon what you learned from browsing the latest art or design exhibition in London, Paris, or New York, but instead clocking your time card into the world of "MVPs"—minimum viable products. Your job as an MVP maker is to reduce your exquisitely grand idea to its most bare-bones possible form, a form that will only vaguely resemble your original vision. Your goal is to create what is perfectly appropriate for a low-cost prototype—which no self-respecting perfectionist could ever tolerate showing the public. But think of it instead as going off and creating what is beautiful and appropriate for the earliest stages of an idea, and then doing the even more important work of finding out what people think about that idea, if even in its minimal form.

You might be wondering, *How could I stoop so low to show my masterpiece before it's finished?* The last thing you want is to experience the utmost embarrassment when a Temple of Design colleague tells you that you've made . . . poop. But the beauty of sharing an incomplete product with others is the opportunity to share it with folks who are unlike yourself. That becomes possible by virtue of network connectivity, and the fact that software can be beamed to most any device in the world. We just need to pop it into the cloud, and then anyone can access it from anywhere.

This cloud is not just storing your product, but also instantly serving your product to all the minds who wish to taste it. By virtue of access to all these people, it becomes possible to learn what they think about your creation. Consider the common practice of posting a feedback box in a prominent place in the company cafeteria—so you can gauge how, say, a new managerial initiative is being received. By the same token, attaching a feedback form to whatever you make on the internet gives you a chance to learn what people think about your MVP. You don't need to guess anymore, because the answers are out there and immediately available to you by gathering reactions from all over the world.

One example of how I've taken to heart this "deliver incomplete and learn first" approach is when I speak at conferences. I can remember when I would spend so much time preparing my TED slides at a meticulous level of detail. But I've since gone in the opposite direction and nowadays show up unprepared. Instead of a carefully designed set of fifty or so visual slides, my new approach is to put my phone number on the screen and ask people to text me any question they'd like me to address. By doing so, I am able to counter my bias to believe that I know what my audience wants. Instead, I'm able to give them exactly what they want instead of simply trying to guess the best I can. I don't believe I would have started giving such presentations if I hadn't immersed myself in Silicon Valley culture. So in addition to keeping the celebrated Japanese essay on beauty "In Praise of Shadows" top of mind as an aesthetic aspiration, I keep a fictional "In Praise of Incomplete" booklet right alongside it in my mind palace.

The beauty of delivering unfinished and incomplete products is that you can always improve them later. There isn't a need to finish them right now—or ever. Look at the internet. Is it finished? Not at all. It keeps evolving with new technologies and advancements coming out practically every day. Consider how, between the eighties and nineties, digital design underwent a shift from print to internet and a new kind of graphic design job emerged: website creator. Unlike designing a poster or book to a finished outcome, a website was designed and delivered—but then it wasn't done. There would always be a need to modify or extend it, because Moore's law was happening. So it fundamentally changed the idea of visual design as something you could deliver to perfection—it was as if the paper that the final designs were being printed upon would deteriorate just at the moment of publication.

Naturally, the Temple of Design railed against Web design for this very reason—because it prevented their high standards of perfection from ever being achieved. So the simple reason why there aren't many websites in museum collections today is that technology is in constant flux, and many art and design experts are waiting for it to . . . stop. The traditionalists are all hoping that, because every trend must come to an end someday, this computational thing will be replaced with the next thing soon. It's natural for them to think and hope this—they don't speak machine, so it's not quite their fault. But there is a healthy avant-garde of machine speakers within the Temple of Design around the world. For example, curator Paola Antonelli dared to collect video games in MoMA's permanent collections, and the art collective Rhizome based at the New Museum in New York has made efforts to offer an open digital archiving solution called Webcorder. Progress is definitely being made, with more to come.

For now, the new kind of product designer learns to become comfortable with the fact that their work might not make it into the history books or be displayed right next to a beautiful chair in an elite museum collection. That's because computational designers are lucky to have the unique opportunity to take what is incomplete and imperfect and make it more perfect by combining their real-time understanding of customer needs with the latest technological advances. They are the ones who get psyched to ship a change to the speedometer in the dashboard of a connected car the day after hearing concerns about its legibility in adverse weather conditions. In the old days, you were expected to know the correct answer, dub it "finished," and simply walk way. Now, you're open to achieving a more perfect understanding rather than a more perfect product. You've come to know that the former is what makes for a truly timeless design in the end because of your willingness to tweak and improve on your formula. You'd rather run a marathon instead of a sprint.

4. An incomplete idea is only a good one if you iterate.

Earlier I referred to the three kinds of style that live under the gilded roof of the Temple of Design: 1) out of style, 2) in style, and 3) making style. Each of these forms of style has its own intrinsic value and raison d'être. There's depth in each of these practices that can easily go overlooked and underappreciated, so let me "double click" on each of them before we go any further in talking about the importance of iteration.

When an idea is "in style," its life span can be extended through variation. You'll recognize this as the fantastic product that at first is available in just a couple of color options—only to later come out in a few varieties. And then later, even more kinds of permutations in color, material, and finishes. Then at some point it meets its end and steps into the "out of style" world. How long the idea remains "in style" depends upon its intrinsic strength, but there's a certain skill to sustaining its longevity that marketers uniquely excel at achieving.

When an idea goes "out of style," it doesn't get to have a decent funeral. It simply disappears and falls into the category of irrelevance that most of us will get to experience at some point in our lives. It was a good idea for its time, and the time has passed. There will be a few stalwarts who wish to hold on to the moment and act as if the world has not changed—but in general they, too, start to fall behind if they cannot keep up with the next trend. Some people happily choose to go "out of style" with a sense of unselfconsciousness that makes them more beloved than shunned, which is embodied by contrarian fashion icons like Iris Apfel, who says, "When you don't dress like everyone else, you don't have to think like everyone else." It's those people who craft the impossible: the idea of the "classic," exempt from the rules of style. Sometimes, that's how even older ideas can end up "making style"—by occupying a renewed status as "retro."

Most aspiring ideas fall into the category of "making style." These may be either a newborn-baby idea, gurgling adorably to everyone, or else an idea from an antiquities collection that just won't die and disappear. It's a category of style that the so-called early adopters are unafraid to welcome and add to their worlds despite knowing that such avant-garde ideas may never transition into the "in style" category and instead take a shortcut into the milieu of "out of style." When you spend time with artists, you'll find that they can be stunningly comfortable when their ideas live in the uncomfortable place of "making style," while fully knowing that their ideas may never go mainstream. It's the more business-oriented designers who wish to see their ideas with mass-market potential become in style, in contrast to the artist-oriented designer who can financially afford to make art that never has to sell.

Now that we can ship incomplete, unfinished products, it's a particularly exciting time for anyone who revels in the act of "making style." Right? There's a reason why Silicon Valley is a particularly exuberant place right now: because radical ideas are given a chance to blast off into the stratosphere—not only to make style, but to make history as well. In the history of technology startups, there has never been a better time to get started than the very moment you are reading this text. *Need computing power?* Borrow time on someone's cloud. *Need people?* Let the cloud find them for you. *Need code?* There are tons of existing services upon which you can get your system bootstrapped and running. *Need money?* Over $100 billion of venture capital funding has been put to work, and there's more to come, with the caveat that investors will expect alienlike, exponential growth rates that often come at a cost to achieving sustainable growth. Nonetheless, there's never been a better time to create a startup, considering all the resources that are now available today.

What is the goal of a startup? It is to become an "endup"—which is a startup that has "ended up" achieving success. The starting part has gotten a lot easier thanks to Moore's law, but subsequently getting traction and ending up as successful as the next Google hasn't proven to be as easy. I was never able to fully appreciate what it takes for an entrepreneur to take a "making style" bet and turn it into a winning "in style" outcome until I spent time in the Temple of Tech in Silicon Valley. It was a pleasure to watch and participate in an unparalleled ecosystem of invention, scale, and possibility that differed vastly from the slower, stilted world of computing research.

STARTUPS	ENDUPS
Want to be something	Already is something
Agile	Stable
Culture is forming	Culture has formed
Have little	Have lots
Have little to lose	Have lots to lose
Try something for the first time	Tried everything and know what works
Unproven	Proven
Do what needs to get done	Clear roles and responsibilities
Flat structure with empowerment	Hierarchical structure with rules
May come and go	Stand the test of time
Heterarchy	Hierarchy

Endup tech companies are often overrun with what is called "technical debt," which has to do with how software gets built. If the software is built quickly and without an eye to the future, then it accrues a kind of debt that becomes evident when a newer piece of software starts to depend upon an older layer of software that wasn't designed to last forever. This happens all the time in the physical world—for instance, when a "startup" bridge is built to span a river in a fast and cheap manner. Before long, this bridge will have become a key piece of infrastructure for a transportation business that hauls goods back and forth over the river. Many other businesses depend upon this bridge to be able to access the talent markets on either side of the river, so many people cross it on their daily commute. At some point, the transportation business wants to use heavier trucking equipment, which the bridge wasn't designed to handle. Plus, the bridge often gets clogged with all those commuters crossing it at rush hour. Putting its existing limitations aside, we can confidently say that the bridge "ended up" a successful project.

Because it would be too disruptive and expensive to upgrade the bridge to allow for more weight or to add a few new lanes to alleviate traffic, the problem is one that its users and operators need to simply deal with. The bridge carries technical debt that could have possibly been resolved if its builders had the time to engineer it to carry a heavier load or to prepare it to easily add extra lanes if its popularity grew. Meanwhile, a little farther down the river there are two separate startups that have received funding to use newer bridge-building technologies, and they're armed with complete knowledge about the flaws of the existing bridge. This might make you feel a bit sorry for the owners of the "endup" bridge, because they will eventually get disrupted by the upstarts. But in the computational world, the constraints are different. Whereas in the physical world it's usually impossible to address any technical debt, in the virtual world it is indeed possible to do so. Why? Because the material is by nature incomplete—it's always subject to change. That doesn't mean it's easy, of course.

The main constraint to addressing technical debt in an end-up company is the people who have made the existing running systems—or, to use my bridge analogy, the bridge builders who are hesitant to modify a bridge that's working fine and well enough. Although software is always malleable and iterable, the question of whether there are the people willing to move forward at the pace of Moore's law is a completely different story. I've worked within, and with, more than a few endups and seen firsthand how the technical debt tends to be managed by valiant, hardworking people who are able to sustain an existing system's performance, slightly enhance it, or slow down how it atrophies. But aiming these same people at ripping out and replacing an existing computational system that's already running at light speed to also risk the health of every other system that's connected to it can (and should) only result in cold stares. Because as software developer Jessica Kerr says,

> The more useful your software is, the more other systems depend on it, the scarier change is. You're risking more than your piece of the world. You need progressive delivery, careful data migrations. Backwards compatibility: twice as many tests, and special cases in the code. You need to design the whole change.

On the other hand, a startup begins with no technical debt out of the gate—but within months can start to accumulate it too. But its overall rapid pace of progress will seem unnatural to the endup watching from farther up the river. At the startup there's a sense of "Yes, we can!" versus "No, don't do that!" because there's so little at stake to lose and there's no single "right way" that has yet taken hold. There's nothing to improve upon, so there's complete freedom to keep "making style" until that one potentially "in style" idea (or, in startup parlance, "product-market fit") makes its appearance on the team's radar. And when rendered as pure software, the opportunity to make constant, steady improvements at a Moorean speed and scale can achieve the outsized gains that companies like Silicon Valley darlings Uber, Pinterest, and Airbnb have managed to attain.

Improvements to a product can mathematically compound, just like money in the Warren Buffett rule. If we make no improvements to a product every day for a year, or 365 days, then it turns out to be:

$$1 \wedge 365 = 1$$

It remains exactly the same and unchanged. But if we improve it 1 percent every day, then it turns out to be:

$$1.01 \wedge 365 = 37.8$$

So constant improvements multiplicatively accrue to becoming roughly thirty-eight times better after one year. Thus, when you have invested in iterating constantly after the launch of an incomplete product, the benefits can be impressive.

Let's consider the alternative scenario, letting the product get progressively worse by 1 percent every day:

$$0.99 \wedge 365 = 0.03$$

It ends up as 3 percent of what you started with on the first day of the year. This decimation of value is an extreme illustration of what happens when you let technical debt go unanswered in a product that is shipped incomplete. A so-so idea with no effort applied to it will, over time, devalue with compounded losses that can cause your original bet to wither away to nothing.

The question of what constitutes a "100 percent ready" idea is debatable because it will depend upon the makeup of the team launching it. Ideally, it is one that can cross the chasm between "making style" and "in style"—by choosing the right raw idea, and then by investing in improving and refining that incomplete idea into one that can constantly be made more complete. There's what Brandon Chu calls the "time value of shipping," which distinguishes between two kinds of projects:

A) Build bigger/better feature → launch slower.
B) Build small feature → launch faster.

If you ship Project B in one month, your customers get to enjoy the remaining eleven months of that year of your evolving product. But if you ship Project A in eleven months and your customers enjoy it for the remaining one month of that year, Project A is not necessarily even twice as impactful as Project B. I have personal experience with attempting to launch my own technology start-up using Plan A and spending an exorbitant amount of my own money, only to realize I should have just chosen Plan B instead. So I urge you to think carefully about the higher value of incompleteness, and the importance of investing in constant iteration while being ruthlessly unsatisfied with the incomplete product that gets pushed out into the world. And keep in mind that the need to iterate quickly should not serve as an excuse to avoid reevaluating your strategy along the way. As my friend Alexis Lloyd likes to say, "Speed and thoughtfulness need to coexist in order to make good things—not just *fast* things."

Lastly, there's a common, inauspicious phrase used when making software products: "launch and leave." It is vaguely similar to a vastly more favorable phrase, "fire and forget," used by the military to describe a missile that can hit its target on its own after it gets launched—it would be nice if that happened all the time, but it does not. "Launch and leave" means pushing out an incomplete idea, never making improvements on it, and accepting the $0.99 \char`^ 365$ downsides of compounding debt. So when you make the shift to the world of incomplete products, be careful that you don't neglect to invest properly in making constant improvements after your idea makes it out into the world. Think of it like a baby who still hasn't learned to walk, so you'll need to care for it so that it matures properly. In fact, it's the antithesis of a "fire and forget" missile—if you launch and leave it, it can rot in ways that will make your computational product go "out of style" before you know it.

5. Emotional value is a need-to-have instead of a nice-to-have.

The special people who make computational products are amongst you, but there are no telltale stains on their clothing, calluses on their fingertips, or any other physical evidence such as you might find with people who've built railroads, airplanes, or any other similarly massive machines. These software people quietly, and invisibly, move millions of numbers into place to construct software machines that move, process, and transform information at high speeds and at extraordinary scales. Although conjuring these computational machines into existence can cause quite a power high, as I've already pointed out, the disconnect with the real world is akin to being a professional gamer or drone pilot, where your immediate reality is vastly different from what surrounds you. And although speaking machine doesn't make you into a machine, it will naturally impact your psyche over time as the non-machine-speakers around you can start to appear blissfully clueless about what's really going on.

When it comes to pushing an idea out to users via the cloud with the kernel of an idea that can be improved over time, it's important to remember who has the most control over making all that magic happen. Who might that be? *You got it.* It's the software developers. Remembering how they can feel is important—and you'll begin to feel what they feel too as you enter their world more deeply. Developers live in the world of infinite loops, ultralarge systems with ultraprecise details, and machinery that's becoming more alive every day. Many of them understand business and design principles, which they may bring to bear on the systems they create. This enables them to ensure that their machines have an "in style" moment and succeed in the marketplace of ideas—think Adobe Photoshop or Gmail, for example. But the basic training of an engineer—as I can personally attest—doesn't usually include business or design training, so a significant gap is prevalent in most technology companies today.

This gap has been exacerbated by the fact that the very nature of business is being transformed by the cloud. In a world where products are no longer finished and incremental improvements are regularly deployed, the relationship with a consumer shifts from buying a product once (owning it) to paying a fee to use it for a set period (renting it). Whereas in the past we might have paid a few thousand dollars for a "finished" piece of software on a CD-ROM, we now pay a few dollars per month for regular access to a cloud-based service. This shift to a recurring revenue model, or subscription business, has many advantages, which include scalability, predictability, and high customer engagement. You're never making a final sale—and you're relying instead on regular income through subscription fees. To illustrate this difference, it's like the distinction between dating and being married: when dating you always want to have your game on, but after getting married you might start taking your mate for granted and get a little lazy. When you're always "dating" your customer, it's critical to constantly please them, especially when they've come to the end of their subscription period and it's time to re-up.

Just a decade ago, pleasing a technical customer was primarily a matter of ensuring that all the machinery worked the way it was supposed to perform. When technically minded people delivered systems for technically minded people, there was a rational exchange of metaphorical bridges to carry cars from point A to point B. There wasn't any need for the bridge to be beautiful, because it just needed to work. Whether the business relationship was recurring or not, tolls are tolls and it made sense to pay for any essential service. But in the new world, where there are so many bridges to choose from and technically sophisticated computational systems are available to everyone, something has changed. Add to this shift the fact that the average consumer is now less often the stereotypical nerd of yore, and instead could be someone's hip over-seventy parent, an iconic reality TV star who's never used a spreadsheet before, or a teenager who sits at the athletes' table at lunch instead of the mathletes' one. A purely functional approach is no longer enough, and instead a richly experiential one has become table stakes because of the relatively new premium standards that have been set by mass market devices and services—think Apple and Instagram.

The minimum viable product (MVP) approach is the minimal or "lean" way to give consumers what they want without it necessarily being a fully realized idea. Given how the cloud works and its unprecedented ability to test incomplete ideas, the MVP approach has become the dominant methodology for pushing ideas out into the world. Although the definition of "viable" is debatable and rightly in the eye of the beholder—because the folks who actually build software systems usually come from an engineering background where "V" will signify being reliable and having as few defects as possible. After all, what good is a bridge if it kicks cars off of it randomly or it can spontaneously collapse? Why bother adorning a bridge with a beautiful floral pattern if it can't withstand the weight of more than one vehicle? Because if the technology doesn't work, then at the most fundamental level the minimal product we're making is going to fail. Engineering viable software systems that are secure, robust, and can be made quickly is not a task to be taken lightly. So "minimally viable" tends to mean prioritizing all available resources on the functional aspects of the product.

Let me linger here a bit longer, because it's important. The engineers are the ones who can "see" the invisible computational world and who are managing a level of complexity that nobody else can see. They're also coping with multiple layers of technical debt that will accumulate as they forge forward, which can easily involve not only the bridge that they're building but also all other bridges that remain interconnected. Only their tired typing fingers can push back against the many dams of chaos and complexity that abound in a world that doesn't reward them for their heroic efforts—no smoke from fire, no stains from dirt, no perspiration from manual labor. So when a business perspective drops into the battleground, or when a design perspective floats in from the periphery, it's important to have empathy for these warriors of cyberspace. In the invisible space of their minds, they are managing an incredible amount of both fine details and massive scales. It can only be helpful to understand them, starting with your new awareness of the computational world. Being able to speak a little machine won't hurt either. When asked how much you speak, you can always coyly respond, "Only a 'bit.'"

A techie might be fine with a rough, purely functional experience, since their tolerance for discomfort is already high to begin with. But the general population has grown high expectations for their apps, so its become important to redefine "viable" as needing to grant a degree of comfort and a modicum of delight. A professional test pilot in an experimental aircraft doesn't need a cozy place to sit, whereas a passenger on a commercial jet will expect a pillow and a soda—preferably the whole can. To make this point clearer in an MVP-ridden world of computational products that are missing creature comforts, I like to use the term "MVLP," where the "L" stands for "lovable." Why? Because it's easy to forget that we're making viable experiences for more than just technical people who prize reliability and efficiency above all. I like to throw in the "L" to remind ourselves that, whether we think we're dating or married to the customer, we still need to play a flirtatious game to keep our relationship solid to remain in business with them.

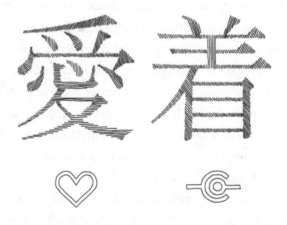

The art and science of design is fundamentally tied to the Japanese philosophy of *aichaku* (AYE-chaw-kooh), literally "love-fit." This design word describes that special connection to something in your environment that fits your life so perfectly that you are immediately bonded to it. Being able to frame the construction of lovable, desirable experiences as proximate to the goals of making scalable, robust computational machinery is no longer just a "nice-to-have," but it's now a "need-to-have." Brokering this connection between the Temple of Tech and the Temple of Design is something I've been doing for most of my lifetime. And for anyone who serves a bridging role in society, you know firsthand that a bridge needs to be stepped on and tested before it can get reliably crossed. *Ouch!* There are often times when I don't know which temple I belong to, and I often get kicked out of one or the other depending upon what I'm doing in a given year. But the awkward balance between these worlds is what has helped me understand and value both perspectives.

A few technology companies have benefitted from the practiced vision of people who understand the psyche of software engineers and who are managed to convert that creative energy into experiences that any human being can love without the Temple of Design's help. So, departing from the usual design success example of Steve Jobs, I like to point to the early work of former Yahoo CEO Marissa Mayer, back when she was VP of search products and user experience at Google. Mayer went in the exact opposite direction of prevailing approaches to visual design on the Web by focusing entirely on the speed it took for information to get from Google's cloud onto your machine. In 2006, she asserted that users "really respond to speed," pointing out how reducing a Google search from 100 kilobytes down to 70 kilobytes led to a material increase in traffic. Place this in contrast to what was at the time the natural inclination to want to deliver a "designed" experience with many full-screen photographic images and other bells and whistles. I attribute the brilliance of Mayer's strategy to Google's newfound position of design excellence, which is rapidly approaching Apple's, even though most celebratory articles largely misunderstand or omit Mayer's seminal contributions.

Mayer took something that engineers could understand and measure, and then used Google's in-house expertise to ruthlessly pursue an experience that could be delivered quickly. This is not unlike what the McDonald brothers achieved when they figured out how to deliver a tasty burger at breathtaking speed with their Speedee Service System. They managed to address a fundamental experiential constraint lying at the foundation of well-designed experiences, which I highlighted as the third law in *The Laws of Simplicity*:

> Savings in time feel like simplicity.

Google's ability to deliver experiences with the right choice of engineering perspective strikes me as truly foundational to their product. Although this early approach often became conflated with a minimalist design approach of "less is more," or with just a nerdly bias, it's way more than that. It's the recognition that the "L" in MVLP needs to be taken seriously at the engineering level from the start, because everything depends upon computation. And so Google's selection of speed as a design attribute stood at the rare intersection of what could be loved by both engineers and non-techies because when a web page loads quickly, it qualitatively feels divine. And once the technical challenge of achieving speed was mastered by Google, they smartly broadened their "L" approach to include nonengineering attributes such as beautiful, highly compressed imagery like what is often featured on their home page.

The lovability of an MVP for nontechnical people can only be made possible by bringing businesspeople and designers early into the inception and construction of digital products of any scale. To slap on a business model to a finished computational machine is no longer a winning strategy; by the same token, spraying design all over it after it's done is a reliable way to lose. An integrated approach is necessary: one that appreciates what developers love to make happen—that is, bridges that don't fall down—in service of consumers who can pay a fair price for an experience that achieves a satisfying *aichaku*-style fit. With the right luck, that teamwork will happen at just the right "in style" moment, when a rising tide of demand can make success more likely for what the Temple of Business knows as the best kind of *aichaku*: product-market fit. Your luck increases with agile methodologies for achieving product-market fit through including your customers in your product development team by inviting them to taste samples along the way.

A common analogy for this fundamental difference in teamwork is to think of an engineering-oriented MVP as delivering a dry cake sample, whereas an MVLP is delivering a cupcake—or a miniature version of a bigger cake with proportionately scaled consideration of all the design details. A cupcake is much easier to construct than a full-size cake and it more faithfully represents the experience of cake as it needs to ultimately present itself. But MVLPs are hard to come by, because there is still a prevailing sentiment and muscle memory associated with the time back when it was impossibly expensive to deliver a cupcake because engineering the cake itself was challenging and few companies could reliably deliver a sturdy, plain, dry cake. A plain cake that you could eat without it exploding into crumbs was the definition of "wow!" and you would love it for its face value as a technical miracle. Today the stakes are higher if you aspire to have mass-market appeal and recognize that the competition can make plain cake just as well as you, so you now need to pursue lovability for a broader range of tastes beyond just the technical ones. It takes conscious effort to compose product development teams with greater diversity of skills beyond technical ones, but incorporating a diversity of life experiences will take the team further. And if all team members can speak even a little machine and have empathy for the invisible and implicit challenges of the computational universe traversed through the fingertips of developers, that's a good start. Then perhaps the combined love for each other and accompanying camaraderie will result in a team that will make what is incomplete even more emotionally fulfilling and successful in the computational era.

Chapter 5

MACHINES CAN BE
INSTRUMENTED

1 // Telemetry gives you a kind of telepathy.

2 // To know you better is to serve you better.

3 // Data science stewards logical interpretations at scale.

4 // Testing your bets is safer than praying you're right.

5 // Autocompletion is inevitable and with no one accountable.

1. Telemetry gives you a kind of telepathy.

On my desk I keep a pile of iron bells that I received from my mother when my parents closed their tofu business. They were attached atop the door frame with just one bent nail, positioned so that when the front screen door would open, you could hear that a customer had arrived. A Japanese-style *noren* (curtain) separated the little shop front from the manufacturing area, so there wasn't a direct line of sight to see customers arriving. And although my parents' tofu shop was small, the loud noises of the industrial machines grinding soybeans made it hard to hear anything. So the little bells' sharp jingling sound afforded us all a kind of low-tech sensing capability to let us know that someone was out front. The bells that sit on my desk have special meaning because they symbolize the power of telemetry even in the mechanical age.

The word "telemetry" was coined in nineteenth-century France, back when telecommunications technology was first emerging. It described the use of an electrical apparatus that transmitted the snow depth from Mont Blanc, the highest mountain in the Alps, to Paris. A key factor that made telemetry especially valuable was that it replaced a person with a sensor, removing the need for a human who gathered data manually on the remote end. The sensor data would be automatically read and electrically transmitted one way to a home base that needed to be within range. Try to imagine what it must have been like to be a scientist in Paris in the late 1800s capable of knowing the snow depth at a location almost four hundred miles away. Telemetry must have felt as magical to the French scientists as the internet first felt when it got started.

Different from the sensor atop Mont Blanc that could only broadcast information, the internet enables high-speed two-way communication—which is an important distinction to make. Two-way communication between two faraway points is subject to errors that will happen from time to time because there's always some kind of imperfection in the system: a faulty wire, a faulty sensor, or even a faulty operator. When computers are talking real fast between each other, they need to ask, "Did you get the message?" And from time to time the other will say, "No. Can you resend it?" As you know, computers operate at high speeds and don't have a problem at all repeating themselves, so they can go back and forth like this until they eventually get it right, because they're constantly acknowledging each other.

Implicit in the idea of an acknowledgment signal, or "handshake," is that the devices explicitly need to talk with each other. It's how they coordinate with 100 percent reliability and don't lose any data. That also means that each device can listen and learn what the other device is up to, because it has to be able to do so in order for it to communicate reliably.

So once you attach a network cable to or enable Wi-Fi on your computing device, it starts to feed on a different source of input than your keystrokes and clicks. Once that connection link is established, the software you're running—if it's instrumented to do so—can send information it might need to share with its manufacturer. This is an important capability to have, just as a flight recorder is important when an airplane makes an emergency landing. For instance, when an app you are running crashes, the software maker can access valuable information about what happened before the program crashed. And by doing so, it can learn how to repair its software so it might not break the same way again. But you will be sharing what you were doing with it just before it gave you problems.

The idea that a software company can learn something about you that you might not want to share when your computer crashes can feel a bit concerning. The reality, however, is that anything you're doing on your computer can be shared with anyone in the world once you enable a network connection. This is especially true whenever you are running a primarily cloud-based system through a web browser—in which case 100 percent of everything you do within a session can be 100 percent telemetered to a remote server. The instant a product is shipped to the cloud and you start using it from your machine, a company can immediately observe how it's used in order to inform future improvements on what is still likely an incomplete design. Most apps or services that you use online—whether you're playing music or editing a shared document—behave this way and are watching what you're doing. This capability gives software companies a kind of telepathy to let them attempt to understand how their customers are using their products. Once you realize this is happening, it's only natural to ask yourself, *How do I turn that off?*

But it is not possible to enjoy the benefits of the conventional internet without this underlying two-way connectivity—there's no way to turn it off. This is the nature of networked software systems: they all have an explicit connection to a system running somewhere outside of your own computer. Your every action can be instrumented and telemetered back to a home base somewhere in the cloud. So instead of being explicitly asked a question about something, like in a pop-up survey, it's possible to simply monitor your behavior and infer what you might be trying to do. The fact that you linger on a particular image longer than the rest can imply you have an interest in it. Or the fact that you have started typing something and then turned around and left the field incomplete indicates hesitation. So even though you haven't communicated anything explicitly, your set of actions can get folded into an inference much as a detective would attempt to piece together clues at a crime scene.

As you can imagine, and perhaps have read in the news, having access to users' thoughts and actions based on their online behavior and data can become invasive. Surprisingly, prior to the 2018 European Union legislation called General Data Protection Regulation (GDPR), there were few impediments to the way technology companies could collect, process, and share data with third parties—all unbeknownst to users. At the time of this writing, the United States has no similar consumer-level legislation, so there are fewer restrictions on what companies can do with your information. And in general, when we click on the long legalese telling us terms of service have changed, we're often giving up some of our rights concerning ownership of the information we're accessing or creating. This is all the more alarming when you consider how it's not just one or two developers at a software company manually analyzing your every action, but instead computational systems that never take lunch breaks or vacations and get down to extreme levels of detail and precision in working to know everything about you.

So if telepathy is a kind of superpower, perhaps we need to look to the code of superheroes as once defined by the late Stan Lee, creator of Spider-Man: "With great power comes great responsibility." It's this responsibility that's coming to the foreground for technology companies, and maybe it's the reason you're still reading this book. The computational approach to creating products is powerful and dangerous—that shouldn't be a surprise to anyone, since power and danger go hand in hand. But the latter concern is frequently overlooked because of what's often labeled as the "arrogance" of today's technology leaders. Yet instead of arrogance, I think of it more as a kind of sheltered privilege that comes from an excess of NIRL (Not In Real Life) time and not enough IRL (In Real Life) time. Having myself lived in the NIRL universe of powerful computational systems that manipulate billions of numbers, I know it can do a number on your ego. Combine the audacity of youth with a concomitant naïveté when it comes to ethics, along with the power of computation—and you have a fantastic recipe for unintended consequences to follow. So as your fluency in speaking machine increases and you learn to accept the gift of telepathy, be careful to constantly widen your aperture of influences to keep the real-world impact of your decisions always in view. Heaven knows how many times I've let myself get comfortable by distancing myself from customers' problems and surrounded myself with people who think similarly to me. Throw in the intoxicating power of harnessing computational telepathy to make yourself believe that you can know everything about anyone, and then the need to consciously exercise higher ethical standards becomes a logical responsibility for everyone who speaks machine.

2. To know you better is to serve you better.

When I lived in Japan for a few years after college, I was always perplexed by how the trains would arrive and depart at the exact times that were printed in the schedules. This was the case for both local trains and regional lines, so I came to realize it wasn't just a Tokyo thing. In contrast, growing up in Seattle, I learned to never trust the timetables for public transportation, and found this to be the case everywhere else—except for Japan. The answer that consistently came back from my Japanese colleagues was matter-of-fact: "Because the Japanese people wouldn't stand for it as customers." At the time, this response came across as a bit snobby. I wondered if they were saying that, as an American, I had lower expectations. My insecurities aside, I understood their attitude of taking care of customers because it was instilled in me at my parents' tofu shop. This desire to care for customers was embodied by a word my father often used: *omotenashi* (oh-moh-tay-nah-shee).

Omotenashi roughly translates to "hospitality," but it means much more than just making someone feel at home. It has to do with how people are greeted and sent off, how they are served, how well you can anticipate their needs and outdo their expectations. In the tofu shop, it meant the rigorous practice of giving customers their bag with two hands and opening the door for them as they were leaving. For my father, it also meant quietly picking the right firmness of tofu to give them if he knew they would be traveling far—so it would not break apart. And although my father would never admit it, it also involved the friendly banter between the customers and my radiant mother from Hawaii with her warm, addictive laughter—which I believe was often the biggest reason they would come back. Yumi always had them leaving with a smile.

Underlying *omotenashi* is having an idea of what the customer wants without asking, so that their needs can be anticipated. There's a famous story that illustrates this best. "The Three Cups of Tea" tells the story of an important sixteenth-century noble warrior who returned from the hunt, evident to anyone that he was deathly thirsty. He was first served tea that was lukewarm and filled to the brim in a big cup, and he quickly consumed it. Wanting more, he was then served tea that was hotter than the last cup, and half the amount. This time he was more relaxed and took longer to enjoy the tea. When he was done and asked for yet another, he was served piping hot tea in a small cup with an exquisite design. With his thirst sated by the first two cups, the warrior could not only fully enjoy the hot tea at the end but also appreciate the beautiful teacup. The tea server, Mitsunari Ishida, was subsequently rewarded by joining the warrior's clan, and later became one of the greatest samurai commanders of that era.

The tale of Ishida's attentiveness to detail is a parable. The idea is not to just serve tea generically, but to consider the kind of tea experience someone might want depending upon their needs. In other words, if Ishida didn't know ahead of time that the warrior was immensely thirsty, he might have just started off with a boiling hot tea in a beautiful cup that would have burned the warrior's tongue. Not only would his thirst not be quenched, but the cup's beauty would be wasted. Or, in blunter terms, if Ishida hadn't made the effort to learn more about the noble warrior's hunting trip—which essentially involved some light spying—then his needs would not have been met so perfectly. When we know about our customers, we have the opportunity to serve them the way they want to be served. But that requires us to be a little nosy, and sometimes a little lucky, to get the information that can tell us how to delight a guest.

You don't have to be in Japan to experience *omotenashi*. It's that moment when your favorite restaurant remembers you by your first name—transforming you from an anonymous customer into one that has "come home" like family. A similar thing happens all the time online when you frequent certain sites that greet you with your first name. Reading a message that says, "Welcome back, John!" feels good at first. But it may feel less good when you visit a completely unrelated site for the first time and it enthusiastically welcomes you by your name. You'd feel similarly awkward if you showed up at a restaurant you'd never been to before and a server you'd never met before is addressing you by first name saying, "John! How's your new job going?" Avoiding this situation—where strangers "know" you to a rightfully uncomfortable degree—is a matter of highest urgency for humanity right now. You'll hear about it in the media regarding our privacy and how to protect ourselves, and it's only natural to want the invasion of our privacy by machines to stop. But to ask a computing device to stop gathering information and to stop sharing it with other devices is like wishing away all the magic in your magic wand.

Computational machinery, by its very nature, can and will be instrumented in some way because this is an intrinsic benefit to the paradigm. The level of instrumentation can vary from capturing your every click and keystroke on a device to capturing your three-dimensional location information on earth at every moment. You may have already heard about how the "cookie" is the basic unit of tracking on the Web. Although they sound completely harmless, cookies are the first sin of the Web while also being one of the reasons internet advertising businesses became so successful during the rise of the internet. Cookies are little pieces of text that any programmer can "park" inside your browser to later access when you come back to it—that way, the browser remembers what you have already visited, and when. It's a handy means for a site to remember where you last left off, and—much like Mitsunari's tea service—it can then strive to make that third cup of tea the best one for you.

Incidentally, this basic technical mechanism of leaving cookies-as-text in your browser's cookie jar also allows services unrelated to the sites you're visiting to park information about you. These are called "third-party cookies," and I recommend you disable them in your browser's settings after reading this. Doing so will let you exercise greater control over your identity and choose who you want to permit knowing things about you—otherwise, it becomes easier for you to appear at that random restaurant that knows all about you even before you've stepped through the door. It's possible to turn off all cookies on your browser, including first-party cookies, but doing so makes the Web much more cumbersome to navigate. Cookies bring convenience, because they mean you don't have to remember a password to a service you've already logged in to—a cookie gets placed on your computer to mark it fully authorized as "logged in" so you don't have to go looking for your password somewhere in your pile of notes. And don't worry, cookies aren't inherently harmful.

For the foreseeable future, you will be constantly trading computational convenience for digital information that you reveal about yourself. The more privacy you give up, the more convenience you get. Said differently, when you share information about yourself, you are guaranteed the pleasure of getting what you want instead of feeling the pain of being served incorrectly. For example, every hotel chain out there is aware that I don't like to stay in a room next to the elevator. In a similar vein, every airline out there is aware that I prefer an aisle seat. Do I mind that they know this about me? Not at all, because it means my desires are more likely to be met. What I do mind is when information about me is disclosed without my permission. But it's hard to know these days when you've given your permission to a company when verbose walls of text pop up to ask you to accept the terms of service before you can get to work. *What did you agree to?*

There are moments when you explicitly opt in to be instrumented and telemetered, like when a site asks you to reveal your location to it. If every layer of technology were to do so similarly, then ultimately you would likely not be able to use the internet as you know it because there are a lot of assumed permissions that we've already handed over. Your new awareness of the complicated nature of the computational universe should make you aware that it's quite possible your internet service provider is storing and selling information about you—in the United States, this is currently fully legal. The same can be said about your telephone network carrier, or the cloud companies, or your favorite apps, or even the physical device you're using—all of them could be independently telemetering and collecting information about you 24/7. The question of knowing how your data is being shared, both willingly and unwillingly, is an emerging dimension of design that

I'm intensely interested in from a computational product perspective. It's a topic that could fill many books, but I'll just leave you with the realization that this is really big. And if you're not fully convinced, check out the American Civil Liberties Union's prescient piece from 2004, "Scary Pizza," which depicts a future when an innocent call to a local pizzeria to order a pizza becomes entwined with the caller's health records and employment history, among other bits of information. The pizzeria ends up charging extra for the caller's attempt to order extra cheese when they're supposed to be on a diet, and the caller is naturally alarmed by how much the pizzeria knows about them.

When you've shared your credit card information with Amazon once, you don't need to reenter it each time you make another purchase. That sounds awesome and feels like magic. When Gmail has processed all your emails and knows how you might respond to a message, it will automatically suggest a response. That sounds magical too. By giving the cloud companies access to all of our information, we enable them to do wondrous things for us and brew the perfect temperature and quality of tea. The only problem is, what happens if hackers break into Amazon and steal your credit card information, or manage to access all your emails from Google? How does it feel? Pretty terrible, right? Is the risk worth it? Absolutely. The way to mitigate the accepted risks is to understand, and to respect, how computational systems work and what can possibly happen when things go awry. To wish away all of the miraculous conveniences that the computational era has introduced would mean that I couldn't easily text my mother a heart emoji at any time or work globally while managing to attend my children's dance performances. Every time technology can save me time or do things better than I could ever do alone, I feel grateful and fulfilled while still cautiously thinking through what I'm getting versus what the cloud is taking from me.

We should be excited about how instrumentation can empower computational products to deliver extreme convenience to consumers by understanding their every want and need. A priori knowledge of customers lets the Ritz-Carlton hotels, for example, provide their legendary service with similar low-tech methods. By noticing that a guest has left a (real) cookie uneaten from their room service meal, they might deliver the next service with a different dessert option by recognizing that isn't their dessert of choice. Having had the experience of staying at a Ritz-Carlton hotel once and enjoying its exceptional facilities and service, I would easily give away all my information to them to receive their *omotenashi*. The question is whether the customer's best interest will be kept in mind vis-à-vis their data. So as you begin to work with telemetered systems and your customers' data, fully embrace the *omotenashi* approach—and treat their data as you would like your own data to be treated. Explicitly knowing the data being shared allows the customer to weigh the compromise they make when losing some of their privacy versus gaining something valuable in return.

3. Data science stewards logical interpretations at scale.

If the average person unlocks their smartphone over one hundred times a day, and there are now billions of handheld computing devices out there, we can figure that the amount of information generated from accessing these devices easily justifies the commonly used term "big data." The massive volume of data generated by all the interactions in the world might make you think that it's impossible for anyone to do anything with all that information. But because you're familiar with how computing machines work, you know that we live in an era where analyzing thousands, millions, billions, trillions, and so forth of data points is entirely feasible—and already under way. A computational machine living in the cloud can collect data at an astonishing level of speed and capacity from a technical perspective, but from an ethical perspective there is an equally great potential for concern. Another cause for alarm is whether or not the data we are streaming into companies will be utilized solely to benefit our happiness and productivity, as we laid out in the previous section. Or will the data be used against us somehow?

How this works technically might be clear to you by now: with the untiring power of a simple loop, it becomes possible to gather an infinite stream of information from any piece of code running in the wild:

```
for( t = 1; < no end condition >; t = t+1 ) {
    < client device shares information to the cloud >
}
```

And from the cloud's perspective—of running processes across all the servers attached to devices around the world—all that's needed is a simple nested loop to go broadly across all servers and then deeply into each individual device.

```
for( server = 1; server < number_of_servers; server =
    server+1 ) {
    for( device = 1; device < number_of_devices; device
        = device+1 ) {
        < sift through all the data gathered from all
            devices attached to each server >
    }
}
```

It's just a matter of hopping over all available servers in the cloud, then stepping through each available device attached to each server. Armed with modern computational analysis methods, the cloud is able to infer the next TV show, book, or outfit you're going to want to purchase—with scientific accuracy and in an amount of time that only gets smaller with each tick of Moore's law. As a reference point, consider that it would take an athletic person roughly sixty-seven years to log a billion walking steps. Now, if the cloud ran using a simple method like I described above, it wouldn't take sixty-seven years to complete, but it certainly wouldn't be the fastest way to get the job done. That's where exceptional software engineers can work their magic to achieve the impossible without waiting for Moorean miracles—thank goodness for the nerds. As recently as a decade ago, being labeled a "tech nerd" was the definition of uncool, but today I can see that stereotype just starting to change as the technology world has broadened. Over the years I've come across many varieties of tech nerds, ranging from library science nerds to targeted-marketing nerds, and more keep coming out of the woodwork. It's important to listen to what Dr. Regina Dugan shared with the TED audience in 2012: "Nerds change the world. Be nice to nerds." And there's a new kind of nerd in town that you'll want to be especially nice to: data science nerds, or, more officially, "data scientists."

Harvard Business Review helped popularize the role of data scientist by declaring it "the sexiest job of the twenty-first century." It gave the following definition:

> Data scientists' most basic, universal skill is the ability to write code. This may be less true in five years' time, when many more people will have the title "data scientist" on their business cards. More enduring will be the need for data scientists to communicate in language that all their stakeholders understand—and to demonstrate the special skills involved in storytelling with data, whether verbally, visually, or—ideally—both.

I recall reading this *HBR* article the same year that I attended Regina Dugan's talk and feeling excited that I had found yet another tribe of nerds just like me, who loved to make sense of data with code. But it took my being a resident in Silicon Valley to fully understand how much data was available to technology companies, and to more earnestly think about the hugged { code block } sitting at the heart of the nested server/device loop on page 147:

```
{ < sift through all the data gathered from all devices
    attached to each server > }
```

You can envision that act of sifting through data as a giant agricultural machine inhaling masses of crops as it canvases a hectare at a time, or instead as a trained cadre of people who carefully pick only the ripe fruits and leave the rest unpicked to properly mature. The former does not bother to discern what it's ingesting because quantity matters more than quality; the latter prioritizes quality over quantity. Clearly, though, a combine harvester effortlessly cruising through a field looks way more impressive than hand pickers. But making sense of it all requires an interpretative skill that we humans are still better at than machines.

That's where data scientists come in. These experts write specialized computer programs that aid in the analysis of all the information gathered. As someone who has long been writing codes to analyze everything from my email to my social media activities to gain greater understanding from available data, I can tell you that it's a lot of fun, even if not a career path for you. If you're wondering, like me, why data science wasn't an available major when you were in college, it's because the field of statistical analysis was only nominally useful, due to two factors: 1) only if there is a lot of data can you get good statistical accuracy, and 2) processing a lot of data by hand is terribly tedious and boring. But computation is the source of the former and the solution to the latter. It seems almost self-serving that computation is the reason why we have so much data while also being the solution to understanding it.

Data science often doesn't mean just writing computer programs. It means throwing out bad data that's collected, bringing stakeholders along so they can make sense of the data, and ultimately deciding on what data to collect in the first place. Statistical terms can be intimidating to hear at first, but it gets easier when you realize that you can classify them as either "descriptive" or "inferential." Descriptive statistics—which you might think of as the easy stuff—involves the various ways to describe a data set in terms of its "mean" or as "standard deviations," in the same way you might label the specific parts of a cow as "brisket" or "sirloin." Inferential statistics involves all the exotica concerned with making inferences from the data; here, you'll encounter mystifying terms like "regression" and "p-values." When you listen in on data science conversations, just remember that an inferential statistical model with a "high R-squared" is good, and if it has "low p-value" then it's exceptionally good.

Even more important than knowing how to do data science is "the need for data scientists to communicate in language that all their stakeholders understand," as that *Harvard Business Review* article so aptly put it. This ability to communicate a conclusion based upon a high-R-squared and low-p-value model requires not only good communication skills, but also an awareness of the tendency to confuse data-backed inferences as indisputable facts. Quantitative information carries a lot of power in an organization of any size—even when it's wrong—so always keep in mind that data itself doesn't produce facts or answers. It only paints a raw picture that requires human insight and interpretation, and thus is never going to be 100 percent correct.

Despite the allure of the seductive graphs, cool certainty calculations, and rational-sounding conclusions that we can achieve quantitatively with data, I've become a fan of another kind of data we can use and learn from: the more qualitative kind. Qualitative data gathering does not get powered by Moore's law but is fed by good old-fashioned listening to real people—which can carry its own level of rigor and science, of course. As user research expert and design legend Erika Hall puts it:

> The best way to assess a functional design is through a combination of quantitative and qualitative methods. The numbers will tell you what's going on, and the individual people will help you understand why it's happening.

If the former is data science, then the latter is data humanism. The science attempts to answer the question, and the humanism makes you ask why it's relevant to people. The science is what makes you run a quantitative study showing that poor people have a greater tendency to feed their children junk food. The humanism is what makes you listen to disadvantaged parents explain that junk food is one of the few things they can afford to show their kids they love them. Having grown up overweight in an underprivileged family, and later educated to think quantitatively at the finest institutions in the world, I'm reminded how all my smarts aren't worth as much as listening carefully to people.

Qualitative insights are fueled by carefully structured conversations with human beings just like yourself—serving as a constant reminder of who this work is ultimately meant to serve. Integrating quantitative and qualitative data is perfectly complementary. Combining "quant and qual" approaches with your team's practiced intuitions will offer you the opportunity to best triangulate conclusions that you can responsibly steward on behalf of the people you serve.

4. Testing your bets is safer than praying you're right.

We've taken the long road to get to the heart of what makes instrumented systems so exceptionally useful: the ability to easily test out an idea. But before we dig deeper into that capability, I wanted you to be at least familiar with how instrumentation happens, how having data about a user can be helpful and harmful, and how there's an entirely new craft emerging to leverage incoming deluges of data. The technology itself is trivial, but its implications are enormous given the world-spanning footprint, speed, and shadow of the computational clouds that connect us all today. By shipping an idea that's instrumented for data collection into the world, it's easy to gauge success or failure by just observing how it actually gets used.

But what's even better than learning whether an idea is good or not, is instead deploying a few distinct variations on the idea simultaneously to learn which is the best direction of the bunch. Maybe a particular aspect of an idea is better than others—this can be revealed by testing variations instead of throwing out an entire idea. It's like designing a fishing lure by attaching three variations of it to the end of your fishing pole and seeing which one catches the most fish. And then, by using the most effective lure as the basis for your next set of variations, it's possible to find the next best lure, and so on. The ability to easily test variations on ideas in this manner reduces the risk of random guesses and can only improve your product design.

Applying our compounding equations from the last chapter to

the question of making successive improvements versus making things worse:

$$1.01 \wedge 365 = 37.8 \text{ versus } 0.99 \wedge 365 = 0.03$$

In the event we got unusually better at guessing at how to improve our system—say, 2 percent a day instead of just 1 percent, then it would be:

$$1.02 \wedge 365 = 1377.4 \text{ [compare with 37.8]}$$

And if we dropped the ball and made things worse by 2 percent a day instead of 1 percent, then:

$$0.98 \wedge 365 = 0.0006 \text{ [compare with 0.03]}$$

This teaches us what happens if we get in the habit of guessing badly too often.

Putting those concerns aside, it's exciting to consider how making a thousandfold improvement over the course of a year is an imaginable possibility. Yet I'd caution you to temper your giddiness over wielding the power of compounding thousandfold gains when aimed at a noble task, like enabling a retired person to afford their weekly grocery bill as opposed to the more sinister prospect of changing a voter's opinion on a candidate. I throw in this cautionary point, as I've tried to do throughout this chapter, to reinforce how the cloud is not just a network of computers but a network connecting human beings. It's certainly easy to forget, so if I sound like I'm preaching, understand that that is not my intent—it's so I can remind myself in the future when I read these words. I find myself forgetting how easy it is to automate bad behaviors.

So when we're running experiments on many computing devices via the cloud, we're running them on real people—not in some invisible, detached computational simulation. For example, in 2014 Facebook shared how it ran an experiment on 689,003 Facebook users in which they were exposed to varying degrees of their friends' negative posts more often than their positive posts, and vice versa. Facebook concluded that it was possible to manipulate a statistically significant number of people's emotions using this method—in other words, the tests showed the potential to fully automate turning a half million people into an angry mob. Now, I doubt that you'll be conducting questionable tests like this on your customers when trying out your ideas, but keep in mind that you absolutely should not do so without their explicit permission and complete understanding.

A more winning example of a successful test is the fundraising experiment run on barackobama.com during Barack Obama's 2008 presidential campaign. Rather than shipping a final website for soliciting donations and then walking away from it, the Obama team ran twenty-four different combinations of buttons and media content to determine that one variation performed 40.6 percent better than the default and could raise $60 million more. The original leaders of this effort for the first Obama campaign went off to start a successful company called Optimizely, which makes it easy for anyone to run such tests. Unsurprisingly, the testing obsession continued in the Obama 2012 fundraising campaign. One example involved testing variations of an email with different subject lines, to learn that one—"The one thing the polls got right . . ."—generated $403,600, whereas another—"I will be outspent"—generated $2,540,866.

When dealing with physical products, the cost of experimenting with variations is still high—although it's starting to fall in certain categories, like when using three-dimensional printing to rapidly fabricate physical prototypes in plastic or metal with increasingly uncanny accuracy. Testing physical variations on people can involve hefty shipping costs as well as costs due to lost time when moving a prototype around the earth, which three-dimensional printing is just starting to address with the ubiquity of printers making it easy to fabricate objects at remote locations. Comparatively, the cost of producing variations of a purely computational product will be significantly lower, even in a world where 3-D printers are everywhere, and especially if the changes involve only the copy or images instead of modifying the actual program logic. Pushing out simple variations can happen instantly and at any scale of reach from just one user or to all users at once.

The best way to learn how to effectively execute these kinds of "split tests" (also called "A/B tests") is freely available in the short 2007 research paper "Practical Guide to Controlled Experiments on the Web: Listen to Your Customers not to the HiPPO." The key point is stated in the paper's title: don't listen to the "HiPPO"—or "highest-paid person's opinion." In other words, let rigorously managed data experiments with your customers guide you to an outcome, instead of letting the boss's opinion overrule your thoughtful work. Keep in mind that a successful variation experiment depends on starting in the right place to begin with—which is usually the choice of the HiPPO. Also remember that having sufficiently divergent variations that make for a worthy test will benefit from tapping into your team's creativity. So it's important to maximize the diversity of opinions around you to get the maximum amount of divergent solutions, which is easily made possible by ensuring there's a diverse team around you with an inclusive culture; otherwise, the quality of your test will suffer.

Testing variations on a basic, core idea works best when you are already comparatively close to success. It's like playing a game of "hot and cold" in search of a hidden object. When you hear "hot"—which means the object is nearby—your best strategy is to take little itsy-bitsy steps and tiptoe around to incrementally make your way to the prize. In contrast, it doesn't make sense to make a sudden, random leap of faith to jump completely out of range, because you were already close to finding the prize. That is, unless there is an even bigger and better prize that nobody knows about waiting just a bold leap away: the proverbial "global maximum" as compared with the "local maximum" you were so fixated on finding. So if you find you're fishing in the wrong ocean and an even bigger bet needs to get made but nobody seems to be listening at the top, then make every effort to overthrow the HiPPO. And when you can't get away with that, Silicon Valley's doctrine says to leave the company and go make your own startup.

Testing ideas that can be instantly delivered to users through the cloud provides an efficient derisking framework when a computational system is thoughtfully instrumented and properly staffed. At the heart of the testing approach and the success of its adopters, like Amazon, Airbnb, and Netflix, lies the simple customer-centric wisdom of Claude Hopkins. Way back in 1923 he stated:

Almost any questions can be answered, cheaply, quickly and finally, by a test campaign. And that's the way to answer them—not by arguments around a table. Go to the court of last resort—the buyers of your product.

Testing with your customers comes with added costs and risks that need to be weighed against the opportunity costs of not making any modifications or attempted improvements. And testing is only really useful when you have a measurable outcome in mind, like nudging a click or driving a sale to compare with a base case you can measure against. Be especially careful of finding yourself in a situation where the default cultural response to any idea becomes "Just test it." On the one hand, this means there's a culture of being open to new ideas; on the other hand, it can suggest a certain laziness has set in to avoid being thoughtful when faced with contradicting opinions. "Just test it" can easily become an excuse for not investing in the informed efforts needed to start off with a really good idea. In other words, testing is real work that can be worth thousandfold improvements and millions of dollars—when you do it right.

5. Autocompletion is inevitable and with no one accountable.

Whereas the Temple of Design pushed us to ship something complete and as close to perfect as possible—epitomizing the high-stakes way of doing business in the last century—the Temple of Tech teaches us to push out something incomplete and instrumented that significantly lowers the stakes to begin with. Then, by subsequently pursuing relentless improvements and experimentation via the cloud, a significant amount of learning follows that helps us to better understand users and serve them best. As a result, tech companies have gotten really good at guessing what you want—whether by the telemetered information we send back to them in real time, or by their machine learning algorithms that make predictions based on our past behavior. We're living in extraordinary times where businesses can do great things for their customers with low risks, ultralow marginal costs, unlimited scales, and at blazing speeds.

As the profound feeling of exhilaration starts to fill you with ambition, there might be a little voice in your head wondering about the future and what role you'll play in all this. Surely you'll be the one worrying about gathering data, analyzing data, and taking action based upon what you learn so that you can test a new hypothesis every month. If lucky, you'll be able to do that maybe once every two weeks—but at some point, if you're running too many tests, it will be difficult to keep track of them. So there will always be a limit to how many rapid improvements can be made on the order of 1.01 ^ 365 = 37.8. But if an "autopilot mode" existed for running rapid tests and improvements that didn't require human intervention, and if it were to realize 0.01 percent improvements every second, we'd get:

```
1.0001 ^ (365 * 24 * 60 * 60 * 60) = Infinity
```

according to Google, but thankfully we have Wolfram Alpha to compute it for us as:

$$2.27085010951860670522948672484157641277497142226720 \ldots \times 10^{82171}$$

which is 2.27 followed by 82,171 zeroes. Wow. That makes my head hurt thinking how hard it is for us human beings to compete against that. Wait. That's not possible, is it? How could a plastic bucket of wires and miniscule electronic parts coveted by 1970s and 1980s nerds make us feel a little . . . insecure and slightly threatened? And how could the infinitely convenient travel-soap-bar-size version of it in our pockets or purses have aided and abetted our potential future irrelevance? After curiosity possibly segues to anger, you're likely to start asking, "Who did this to me? The tech companies? It must be! And now how do I make them pay for it?"

It's logical to blame the technology industry for the universe of computational machinery that permeates every aspect of our lives today. But it's not their fault alone—it's your fault too, because you were oblivious to what the Zuckerbergs of the world have long understood as fluent speakers of machine. If computation was something we could all see and sense in the physical world, we would have long gotten used to a new set of speed limit signs, speedometers, and the like. Consider how the first car to break the sixty-mile-per-hour limit was a torpedo-shaped electric vehicle called La Jamais Contente—The Never Satisfied—invented by Belgian engineers and tested in France in 1899. If La Jamais Contente were to have evolved at Moore's law standards, it would have broken the speed of light by 1925. Surely, if by 1910 we'd noticed that the Belgians or French were zipping about on the surface of the earth past the speed of sound, then somebody would have done something about it. The changes that have quietly occurred in computation were invisible for such a long time that we're only noticing the impact today in politics, media, and business.

Our biggest opportunity—and also our biggest problem—has to do with the data that's being gathered not only right now, but also from our collective pasts. Imagine the tech industry's cloud system floating above us all as an endless array of eternally desiccated sponges that hydrate and get engorged with everything that can be gathered about what we do, where we are, and what we think. On the one hand, it can seem ominous, dark, and terrifying if you can't fathom what it's capable of doing—which is what most people are starting to feel. On the other hand, we can feel gratitude that data and ethics experts have long been thinking through many of the issues so that we can start to distinguish between ethical and unethical sources of data collection. The computational world was long invisible, but it's become visible, and for us to fully trust it we need it to be more transparent and understandable. We're no longer going to be OK with opaque, black boxes that we can neither look inside nor interrogate when they misbehave.

We're still in the infancy of understanding what it means when the data owned by a credit card company can be completely anonymized and sold to a company like Google, to later be deanonymized by simply matching the credit card transactions to location information that's either broadcast by your smartphone or tagged by information you might share about yourself online. We're still at the point of infancy in grappling with what has been going on at Moorean speeds that has long gone unnoticed. We can bet that if the equivalent of La Jamais Contente zooming past the speed of light had been noticed in the late nineties, the way that we at the MIT Media Lab palpably could recognize back then, the world might be a different place today. Instead, the entire physical world is coming to realize that there's an invisible computational world that the tech companies fully control. Still, we should give the benefit of the doubt to the many nerds inventing these technologies—who aren't doing so with malicious intent but are instead driven primarily by curiosity—and we should definitely punish the bad actors because there's definitely an unhealthy share of unacceptable nerds out there. Because there is a gradual awakening for techies that's begun to take hold—for while they are brilliantly composing lines of code, they're starting to wonder what happens when their incomplete computational systems learn to "autocomplete" themselves someday soon. Even the cleverest of automators are not safe from being cleverly automated too. Luckily, that has started to make them all a little nervous and willing to rethink what they've made happen in the world.

There are roughly five phases of evolution in the software product industry that culminate in an end state where AI will take care of everything because it will be all-knowing. We've made it through the first three phases and are currently in the middle of the fourth, where human intelligence is blended with computational intelligence in a kind of hybrid—thus dubbed the Centaur era by Nicky Case—to signify the melding of a human intellect with the superior physical stamina of a horse. We know today that while we make computers smarter they are augmenting our smartness too, so instead of beating them we choose to join them.

1. Shrink-wrapped Boxes: We shipped software in boxes with tamperproof plastic and shipped updates by the same method.
2. Shrink-wrap + Download: We made boxed software optionally downloadable online, and we made updates available online.
3. Software as a Service (SaaS): We moved software into the cloud as a service, and human teams labor to continuously update it.
4. WE ARE HERE—> SaaS by Centaurs: We run software in the cloud that's constantly being improved by human teams that collaborate with light AIs.
5. A New Beginning: We will be using software that evolves more rapidly than ever because it's powered by know-it-all heavy AIs.

If you're nervously looking about to spot a centaur somewhere in your field of view, their fruits already live in the kind of computational machines that you access today from your smartphone or other devices. They've been busy at work on your recent shopping trip to Amazon, and have completely rearranged your shopping experience with *omotenashi* to accommodate your welcome visit back. Meanwhile, they've done the same for millions of other Amazon customers simultaneously. The same can be said for your recent visit to any other shopping, search, news, or video service, because computational *omotenashi* is increasingly available to serve you in the way you wish to be served. The easiest way to confirm this is to look at what your friends see on their screens—or even better, the screens of people who are completely different from you in age, culture, beliefs, gender, or any other traits to distinguish them from your immediate circle of friends.

Media Lab cofounder Nicholas Negroponte famously posited this future of fully tailored customer care back in the 1970s with his "Daily Me" concept of a newspaper that was completely customized to just what interested you. It's now a reality in every facet of our digital lives, because all the centaurs have been working hard to automate and constantly improve the hedonic bubbles surrounding our brains. On the one hand we're experiencing the best moments of our lives played and replayed back to us to enjoy. And on the other hand we're experiencing a kind of cognitive gratification that can ultimately limit the range of our individual and collective growth as a species. Do we blame Negroponte? Do we blame the centaurs? Or do we blame the technology companies? As elected leaders begin to speak machine, they will seek to intervene on behalf of the common good, as we see already happening. But given the inertia of governmental processes, and the antiquated mind-sets of many lawmakers it seems unlikely legislation will ever be passed that can fully address the impacts of the one law that's not really a law: Moore's.

If you're reading this book in print form, then you know it's not instrumented and you can feel completely comfortable that I have no idea which pages you have ignored, so I won't be offended. But if you're reading it digitally, then . . . you got it: I know what you've been reading—but I still won't be offended if you've skipped anything. Because you're someone who's more curious than afraid of what's to come in the future, and I wrote this book for you: the new centaur. (Is that extra set of furry legs bothering you yet?) But the way of the centaur has its own set of problems to deal with, given the sense of superiority a magical being is prone to feeling. Case in point, even when we've become split-testing experts, we're likely to only succeed a third of the time, fail a third of the time, and have no impact otherwise. When we write that out using our equation of compound improvements and compound failures, it becomes:

$$1.01 \wedge (365/3) * 0.99 \wedge (365/3) * 1.0 \wedge (365/3) = 0.987906 \ldots$$

That's 0.98, which is essentially 1, which means there is no net cumulative improvement when we play this game—and furthermore, if we increase the frequency of updates out to every second, the result of the equation is zero! I'll ask you to not read too much into these simplistic calculations, but I'd love for you to have fun with me in wondering how we human beings can nudge improvements in bold ways that the computer will not be able to figure out on its own. Take pride and responsibility in knowing that our interventions have the potential to prevent the computer from dramatically blazing its way down to . . . zero.

Back when I was at MIT, I declared myself a "humanist technologist" without knowing what it truly meant, primarily motivated by the thinking that the default motto of the technologist was that more technology is better. Technologists create progress. Progress is about things happening. And we wouldn't want a world where nothing happened at all. But in time I came to see the difference between the two terms:

> Technologist = I do, because I can.
> Humanist = I do, because I care.

With just the substitution of two letters for one, "can" turns to "care" and technology becomes humanized. I believe that you care. Because you care, you are now ready to proceed to the final chapter to learn about the one challenge that AI is not equipped to manage, showing that we good ole humans are relevant—and are emphatically needed. The grand task of addressing the impending automation of an imbalanced society is upon us right now, just as we are on the precipice of the Singularity, and it is not too late for us to do something about it. If and when computers fully outpace the intelligence of the entire human race, there will always be certain things that machines will not be able to beat us at doing, and it's our job as humans to figure them out. By knowing the world of the machine, you will need to hold yourself accountable to what we as humans have managed to create together—and now we need to re-create together before it's too late. A computational product will be subject to immediate improvements driven by the data collected about us, and with AI's rise as the sole force in driving future iterations without our active and explicit input, we need to do something about it.

MACHINES AUTOMATE
IMBALANCE

1 // The tech industry has an ongoing
tendency toward exclusion.

2 // Big data conclusions need thick data
connections to real people.

3 // We should expect artificial intelligence
to be just as dumb as we are.

4 // Open source is a computational means
to engineer equity.

5 // Mind the humans.

1. The tech industry has an ongoing tendency toward exclusion.

Given that the technology industry has long been fed by the technology education industry, where I'm originally from, I had long assumed that the matters we addressed at the source would make their way downstream. Just a few years after I had joined the MIT faculty, then president Charles M. Vest released a statement following an official report about the women faculty at the MIT School of Science in 1999:

> I learned two particularly important lessons from this report and from discussions while it was being crafted. First, I have always believed that contemporary gender discrimination within universities is part reality and part perception. True, but I now understand that reality is by far the greater part of the balance. Second, I, like most of my male colleagues, believe that we are highly supportive of our junior women faculty members. This also is true. They generally are content and well supported in many, though not all dimensions. However, I sat bolt upright in my chair when a senior woman, who has felt unfairly treated for some time, said "I also felt very positive when I was young."

I naively thought that with this statement made public by MIT, and subsequently reported by *The New York Times*, the great injustice of gender discrimination was henceforth officially banished from the technology world. So I was more than flabbergasted when

I arrived in Silicon Valley over a decade later and was introduced to a room full of the "top UX designers in Silicon Valley," only to find two women present. By the time I left Silicon Valley, the situation had improved to fifty-fifty for any gathering I organized or participated in as a speaker, because I found it to be the most effective way to improve the overall quality of an event for everyone attending. So it only felt logical to me.

As I began to dig deeper into the statistics for tech, I became concerned. I learned that there was only 21 percent representation by women in tech, whereas the overall population of women in the United States is roughly 50 percent—an obvious imbalance. In 2014, the US Equal Employment Opportunity Commission reported that the high-tech sector employed 7.4 percent African Americans, 8 percent Hispanics, and 14 percent Asian Americans, whereas the average overall representation in the private sector was 14.4 percent for African Americans, 13.9 percent for Hispanics, and 5.8 percent for Asian Americans. As an Asian American, I couldn't help but notice a study by the Center for Employment Equity reporting that, despite the relatively high proportion of Asian Americans in tech, managerial and executive jobs are more likely occupied by white men—or "pale males," as the new lingo goes. Outside the United States (China, for example) there's a general inclination toward males in the tech industry as well—which indicates that it's not just a problem limited to the paleness of one's skin.

Even more concerning than the simple disproportion in these figures is how such imbalances affect the quality of life for workers outside of the majority in the tech industry at all levels. The Kapor Center for Social Impact researched the main reasons why these folks leave the tech industry, citing discrimination, bullying, sexual harassment, and racism as the top causes. The study also found that women and people of color were the most likely to experience harassment and be passed over for promotions. To connect back to my MIT story, in hindsight I can see that MIT and most universities across the world never aligned with Dr. Vest's ideal "reset." And so the system that has fed much of tech with engineering talent—which has a gender imbalance to begin with—has simply done what all systems do when they're biased in a certain direction. They just continue along the default path. And leaders who are awake to those lost opportunities to create higher-quality work environments can choose to do something about it or just let it be.

From a systems perspective, we can predict that an imbalance in the tech industry of this magnitude will likely perpetuate without self-correction. Tech companies need to run at full speed to keep up with Moorean time scales, which fosters the pressure to optimize for "culture fit" among potential hires—meaning people who are "just like us." That way, a new person will take less time to onboard (because they are "like us"), create less day-to-day friction (because they are "like us"), and follow the boss (because they are "like boss"). And these people, in turn, will hire more people like themselves—unless there are explicit system-wide interventions and incentives or penalties that can break this cycle. Whether it's the friends you went to college with who have similar tastes, or the people in your neighborhood who moved there for similar reasons, or your professional circle that's already sorted itself for maximal camaraderie, our tendency is to reduce friction and choose sameness over difference.

So it shouldn't surprise anyone that the tech industry is filled with people who are more likely to think alike and come from similar backgrounds, because the need to move fast will always outweigh a slower, considered approach. But when there is a "we" that defines us, there is a corresponding category of "not like us" that is naturally excluded because "they" think differently and will slow us down. The Temple of Tech is no different from the Temple of Finance or any other specialized profession that seeks to foster its own culture. The boundaries of any temple will nurture safe cultures of like-minded people who prefer to avoid the friction they might feel whenever they're not with their own tribe. The difference is that, although we should care about inclusivity in any sphere, the techies exert disproportionate influence as they operate at a whole different Moorean speed and scale.

An "oops" by a professional in any company can negatively impact many people, but an "oops" in a computational system can impact all connected customers within a few milliseconds after the tap of a keystroke. When the biases of a business monoculture lie at the foundation of an "oops"—like in an emailed reply-to-all from Finance that subtly puts down anyone who doesn't know what EBITDA means—that can be unfortunate. But when the biases of a tech product team are deployed to millions of users simultaneously—such as copy in an on-screen button label that will read as insensitive to non–pale male customers—that takes it to the next level. Fortunately, the feedback loop of social media is swift and relentless when aimed at companies' missteps, but there are even worse "oops" that can sit deeper within a company's culture. For example, when an internal hiring tool was programmed by an Amazon team of majority pale male AI experts who'd leveraged past data from hiring decisions by likely majority pale male managers, the computational systems demerited résumés that mentioned attending women's colleges or used the word "women's." So what can easily be pointed to as a "computer program error" needs to be considered as more of a "culture error" if we are to truly prioritize accountability.

An imbalanced system will produce imbalanced outcomes. When applying that thought to the tech industry, we can expect imbalanced products to be produced for the foreseeable future. With the players in the Temple of Tech running at computational speeds and scale, we can expect the velocity and level of imbalance to be unparalleled—and eventually fully automated. Beyond the social implications of why that's not a good thing in terms of equity and justice, from an organizational perspective it represents a suboptimal path for achieving breakthrough innovations. A culture of sameness, devoid of catalysts for innovation, is a losing strategy for a business to achieve outstanding growth. It's also a source of risk to your business's ongoing performance when you ship an insensitive "oops" in your product that could have been avoided if the team had been more inclusive of diverse backgrounds and viewpoints in the first place. A work environment where no one is afraid to speak up for fear of having a different point of view is how costly mistakes are most quickly avoided. But you need different kinds of people to hear different kinds of perspectives.

Sara Wachter-Boettcher's landmark book *Technically Wrong* documents the many ways the technology industry has unconsciously let the biases of its primarily pale, cisgender male culture impact its products. The result is everything from menstruation apps that refer to users as "girls" to shopping apps that push notifications to women to shop for a Valentine's Day gift to delight "him." Or when a popular social media company released a real-time image filter to add slant eyes to any face to approximate an Asian caricature—just months after it had released a real-time image filter to darken a light-skinned face to look black—such unacceptable mistakes result in PR backlash that, while costly, may not result in the hiring investments that are necessary to make better product decisions. So this can be expected to continue as a natural outcome of the imbalance inherent to the tech industry, which also goes to the heart of how startups are funded, managed, and overseen by their boards—making those compromises ripe for disrupting.

Smartly, there is a wave of pragmatic business leaders in tech who see new opportunities in adopting more inclusive approaches to creating their products. They know that failing to serve the broadest possible range of customers represents a kind of old-world ignorance—and leads to lost business opportunities. Instead, they're actively working to diversify their company cultures to serve their customers better by, for one thing, addressing the gender pay gap in the tech industry. There's interest in moving from the narrow-mindedness of the "culture fit" approach to instead valuing differences as "culture add"—that is, bringing new voices and ways of thinking into an organization as a positive asset. For all of Google's diversity and inclusion challenges—as illustrated by the firing of an employee for his internal antidiversity manifesto memo—the company has been steadily investing in a promising area called "product inclusion." Headed by business leader Annie Jean-Baptiste, the initiative takes as its central thesis that diverse teams make better products. As a result, Google is examining everything from supplier diversity for sourcing of supplies and equipment to an "inclusive images competition" to populate their image databases with more diverse representation. We will see more of these kinds of sanguine efforts as consumers demand higher standards not only in the quality of their products and the ethics of how they get made, but also the character of the company they do business or share data with.

The challenge of righting the imbalance in the tech industry presents an immediate opportunity for new business growth and disruptive innovation. Accepting this challenge can feel at times daunting, even impossible, when considering the deep roots in the inequality of tech education, and even deeper in the wealth inequality that spans multiple generations in our country. Yet I'm increasingly optimistic, because I believe that the most leverageable starting point is for the power of computation to be comprehensible to more people—specifically people who were previously unaware of its implications because it lives in an invisible universe. It wasn't your fault that you didn't know much about it, as you couldn't have noticed it in the first place. But now you know it's there, and everywhere. And when you consider the endless, untapped opportunities at hand whenever we cross national, race, gender, cultural, religious, age, and socioeconomic boundaries in thinking about future products and services, there's undoubtedly plenty of innovations to be born. This is called "greenfield" or "white space" in the jargon of my businessy colleagues—a space where there's still nascent development or prior ownership, so newcomers still have a chance to enter and grow their influence and build new business opportunities.

Imagine rebooting tech education by following the example of President Maria Klawe at Harvey Mudd College, which has achieved gender parity in its computer science program. Or consider how Advanced Placement computer science exams have seen record increases in female, black, and Latino students, in part by shifting the emphasis away from pure computer programming toward instead addressing meaningful problems through code. Or consider the incredibly diverse learning ecosystem of Word-Press, with its network of open-source volunteers of all ages who informally teach each other how to code and use digital technologies to gain practical skills to put food on their families' tables. All it takes is for the sameness of tech to tap into the full diversity of humanity, and we can potentially rid it of the imbalances that have been created at the speed of Moore's law. Sound impossible? Absolutely. But computation enables the impossible, and if we fully harness its capabilities, we humans can re-create the Temple of Tech as one that is welcome to all.

2. Big data conclusions need thick data connections to real people.

Given that we can easily deploy computational products that are incomplete and instrumented, we have the opportunity to get back in return a large amount of data to determine how to modify and improve those products. Success will usually look like harvesting lots of user data as the best means to improve the statistical accuracy of your data-driven conclusions. So with computational products, it's easy to become biased toward broadly observing the behavior of thousands of users instead of consciously investing time to delve deeply into just a few individuals' experiences with your products. Why? The simple reason is that it's so much easier (and thus cheaper) to take the instrumented approach to studying aggregate behavior, due to the computational power available to us today. It also makes you look super smart when you can rattle off a convincing, scientific-sounding factoid like "7.2 percent versus 1.2 percent." By contrast, it's much more low-tech (and thus more expensive) to study individual people's behavior through methods developed by anthropologists—in essence, ethnography. It's a positive sign of customer empathy to share insights after spending time with that one nontechie customer named James who is having a few challenges with your product.

Here's the problem: the "scientific" response given as numerical data will usually get the majority of affirming nods compared with the stories of James's challenges and the roadblocks they're currently facing. That's because a quantified viewpoint appears like fact—an important signal extracted from the noise—while a qualified viewpoint appears like a noisy customer who just doesn't "get it" and can be discounted because he falls outside the 7.2 percent. In actuality neither the aggregate data nor the individual's story constitutes fact, because both contexts involve people. Human beings are by nature unpredictable, so anything involving predictions of human behavior is ultimately going to be a guess. One kind of guess uses quantitative data and the other uses qualitative data. We pay big money for high-quality guesses as one of the means to lower risk in decision making, but no guess can give a 100 percent guarantee of success. That's why it's a guess, not a fact. And the best way to guess better, as any investor knows all too well, is to create a portfolio of bets so that all the casino chips aren't placed on only one of the guesses.

It's gotten so easy to harvest quantitative data in the computational era that part of the challenge today will be to rip techie folks away from their standing desks with giant monitors and copious snacks to do an old-fashioned, face-to-face customer visit. This is especially difficult since effortless access to quantitative data is a primary benefit and de facto outcome of the computational era, so to folks who live every day in the future, it may feel like going in the opposite direction. Furthermore, if it only costs five dollars a month to gather and analyze millions of your online customers using your products, it can seem unnecessarily expensive and inefficient to make the investment in working one-on-one with a customer, which can cost hundreds of dollars per month. To draw on another finance industry analogy, the best investors will not only carefully analyze the funds they are engaged with, but they will fly out and do a site visit to the fund managers for that extra bit of due diligence as an investing professional. So if the due diligence of the investing world sets the highest standards, then talking with a real customer from time to time makes good business sense.

This is the lesson of good ethnography: to understand a cultural phenomenon, you need to get as close as possible to "first-source" information, instead of relying on second- or thirdhand information. Furthermore, to truly understand first-source information, you need to invest time in knowing and understanding the cultural context that surrounds it. Cultural anthropologist Clifford Geertz defined the ultimate goal of ethnography as "thick description," as opposed to "thin description." Thin description focuses merely on the superficial details, whereas thick description goes much deeper than immediate observations and attempts to capture the many layers beneath just the surface.

For instance, from my own experience working on WordPress products, I know it's not uncommon to hear a thin description like, "Ninety percent of people spend most of their time checking their blog's viewing stats," which leads to the conclusion that it's an important feature that needs to be improved. But such an analysis is quickly disrupted by a thick description from a user telling you that their blog's viewing stats are the first page they land on in WordPress, and since the stats page shows zero views, they're not motivated to continue using their blog. So the problem isn't about improving the stats page, the problem is enabling a blogger to write content that garners views and builds a readership base. It's easy to let impressive-sounding aggregate data present a persuasive case for action that can miss an underlying, bigger problem. So when presented with quantitative data, it's important to demand what tech ethnographer Tricia Wang calls "thick data," in contrast with "big data." Gathering thick data takes time, and interpreting it well can take even longer. You need to marinate in the thick data that you gather to fully capture the many contexts of your fellow human beings, or else there will be little benefit from your added investment. The allure and ease of quantitatively processing big data will constantly pull you away from the time commitments required to comprehend thick data.

As a fellow busy person, I confess that I like to hide behind my computer screen and sit in my comfy task chair, because I can efficiently get a lot of work done and my rhythm doesn't have to get disrupted by upending my surroundings. But ever since I started to actively work face-to-face with customers—as inspired by Intuit founder Scott Cook's habit of "going home with the customer," watching them install and run his software system, a practice he began back when he first launched Quicken—I now fully realize that the time is well worth the investment. It raises the stakes in your work of serving customers. It can be a terribly uncomfortable thing to do, because you will quickly know when you've let down a fellow human being with the decisions you've made in your product. And when gathering thick data on your own, be careful of how easily you can become biased that your one customer's problems are every customer's problems. By now you've embraced imperfection, so just go for it.

The one piece of advice I'd give for "going thick" is to try not to focus on the specific problems your customer is facing with your system. Instead, keep in mind their overall goal for wanting to work with you in the first place. For example, I recall how in the nineties Japanese copy machine makers were designing elaborate user interfaces to manage paper jams, only to be blindsided by organizations going paperless as a better way to share information. I liken this to what we often encounter in customer support as managing a lot of "flat tire" situations, and then we immediately want to spend all our time to make a tire that can't go flat—or more often getting really good at fixing the flat tire. Meanwhile, we can forget to think about where the customer was going in the first place—asking instead, "What was their destination, and what hopes and dreams were associated with it?" By starting from the motivation question as the driving force behind thick data, you'll remain more strategic as you immerse yourself in firsthand information. Remember that you're looking for subtle, human details that are impossible to capture with charts and numbers, so try to rely on your ability to smell and feel a situation. Be what AI cannot do.

I learned a version of this lesson as an undergraduate researcher at the MIT AI lab working for a visiting engineer from Digital Equipment Corporation, or DEC as it was affectionately called until it vanished like many early computing companies. She told me an unforgettable story about how a major soup company had invested a fortune in creating an "expert system" (the first generation of AI) to make soup in their factories just like the human operators. The soup company's problem was that their best factory operators were all getting older and they weren't sure how to deal with them all eventually retiring. So the soup-making experts were carefully observed, and all their actions and ways of thinking were then encoded as IF-THEN rules. The day finally came for the factory to fire up the *au levain* AI system and make some soup. But the results were disappointing—the soup tasted terrible, in fact. Now, at the time I was a big fan of expert systems and was shocked to hear about this failure, so of course I asked the visiting engineer if they'd figured it out. "It was really quite simple and funny," she said. "They asked one of the old guys to explain why the soup tasted bad. He stepped forward, leaned over the soup bowl, and sniffed it a few times loudly. His response was, 'It smells bad.'" I love this example because it still holds true today. Indeed, complex systems have many intangible aspects that are easily dismissed with even the most computationally advanced techniques. Being human is still pretty cool, so f*** the AIs.

So with your nose pointed forward into the future, keep in mind the three traps that have always gotten me in trouble. You'll bump into them as you get more computationally adept and more design-y opinionated, and also simply by getting older like me you might start to believe your own blah-blah-blah. I'll be brief so you can rush ahead to the end of the book as I know you're almost there!

1. Thinking like a classical engineer, and believing there's got to be only one way to build it right. Henry Ford believed, like a good engineer, that everyone would want a Model T available in only one simple, pragmatic configuration, and painted black. Alfred P. Sloan at General Motors believed that there should be many kinds of cars for many kinds of people to give them what they wanted. Ford lost and GM won by having a better nose.

2. Thinking like a classical designer, and believing your solution is the one that all will bow down to and adapt. The standards that elite institutions uphold as the cultural compass for the classical design world are underwritten by subjective decisions and invisible wealth networks that facilitate what gets remembered versus forgotten. The Temple of Design narrative of the "genius designer" is not a reliable pathway to success—it's seductive, but it's stupid. Use this nose less.

3. Thinking like a senior leader, and believing that what worked well in the past is obviously applicable yet again. I've trained to catch myself when I say, "Back when I was at X, I did Y, and this problem we're facing is the exact same thing again. I know how to solve this one. Follow me!" I stop myself here. That's because I know we live in the computational era, where we can't expect what worked ten years ago to apply right now. This is what

entrepreneur Barry O'Reilly calls having the rigor to "unlearn" from past successes, or you will miss new ones. So when in doubt, go get a brand new nose.

We can address the many imbalances that have seeped into the technology universe by focusing on the human element of our work. Computational machines are master copycats and are powered by the quantitative data at their foundations. So we'll need to pay attention to our completely normal "human nature" of relying on our usual biases, aka "wisdom." The computational era will easily turn out poorly for all of us if we don't balance out all of our quantitative data with more qualitative data. So start now and broaden your data portfolio. Invest like a smarter-than-average boss. And gather as many observations from people who are unlike yourself—because triangulation works best when you have the most diverse set of sources with which to tune and retune your data-informed guesses. Rather than just trying to protect your nose from a thick and occasionally unpleasant smell, we live in a time that requires your full sensorial attention and your inquisitive curiosity. Go thick. Smell hard.

3. We should expect artificial intelligence to be just as dumb as we are.

When the cofounder of Google talks passionately about his fear of what AI can bring, it isn't just a ploy to boost Google's share price. It's because those who have been (literally) plugged in to the power of Moore's law know something about its impact that the general population doesn't know. When Harvard scholar Jill Lepore says this about the transformation of US politics:

> Identity politics is market research, which has been driving American politics since the 1930s. What platforms like Facebook have done is automate it.

the key word is "automate"—because machines run loops, machines get large, machines are living. It's not like turning a steel crank by hand to make a plastic toy figurine move about. It's like pressing a button and watching the toy get up, wave at you, and then start answering all your emails for the rest of your life. When a young person says on camera that Cambridge Analytica and Facebook together were able to sway the 2016 US election, we look at them and think, *No way.* Because there can't be that many human workers in the world who could process millions of pieces of information at a low enough cost. But you're aware of computation now, and understand that the present is significantly different than even the recent past (just over a year ago).

Automation in Moorean terms is very different from simple machines that wash our clothes or vacuums that scoot about our floors picking up dirt. It's the Moorean-scale processing network that spans every aspect of our lives that carries the sum total of our past data histories. That thought quickly moves from wonder to concern when we consider how all that data is laden with biases, in some cases spanning centuries. What happens as a result? We get crime prediction algorithms that tell us where crime will happen, so officers are sent to neighborhoods where crime has historically been high—that is, underprivileged neighborhoods. And we get crime-sentencing algorithms like COMPAS, which are likely to be harsher on black defendants because they are based on past sentencing data and biases. When asked about AI and its ramifications, comedian D. L. Hughley optimistically replied: "You can't teach machines racism." Unfortunately, his assessment is incorrect, because AI has already learned about racism—from us.

Let's recall again how the new form of artificial intelligence differs from the way it was engineered in the past. Back in the day, we would define different IF-THEN patterns and mathematical formulas, like in a Microsoft Excel spreadsheet, to describe the relation between inputs and outputs. When the inference was wrong, we'd look at the IF-THEN logic that we'd encoded to see if we were missing something and/or we'd look at the mathematical formulas to see if an extra adjustment was needed. But in the new world of machine intelligence, you pour data into the neural networks and then a magic black box gets created: you give it some inputs, and outputs magically appear. You've made a machine that is intelligent without having explicitly written any program per se. And when you can feed it with tons of data, you can take a quantum leap in terms of what machine intelligence can do with newer deep-learning algorithms—and the results get significantly better with the availability of more and more data.

Machine learning feeds off the past. So if it hasn't happened before, it can't happen in the future—which is why if we keep perpetuating the same behavior, AI will ultimately automate and amplify existing trends and biases. In other words, if AI's masters are bad, then AI will be bad. But when systems are largely running on autopilot like the new AIs, will media backlash lead to an "oops" being attributed to AI error rather than human error? We must never forget that everything is a human error, and when humans start to correct those errors, the machines are more likely to observe us and learn from us too. But they're unlikely to make those corrections on their own unless they've been exposed to examples set by a sufficient number of humans who can provide the right corrective behavioral data to rebalance their numerical brains.

In our machine intelligence era, an apt analogy is how children inevitably copy what their parents do. Oftentimes they can't help growing up to become like their parents—no matter how hard they try otherwise. While in the old days solving a complex problem by writing a computer program could take months or years, now machine intelligence can rapidly conjure an equivalent computational machine when just fed with past data by which it can model a past behavior. Automating a past outcome can happen instantly and with increasingly less human intervention. So rather than imagining that all the data we generate gets printed out onto pieces of paper in a giant room somewhere at Google, with a staff of twenty running around trying to cross-reference all the information, think instead of how machines run loops, get large, and are living. The logical outcome of that computational power—a rising army of billions of zombie automatons—will tirelessly absorb all the information we generate and exponentially improve at copying us. The AIs are not to blame when they do bad things. We will be the ones to blame for what they do in the service of us.

With revelations that systems like Facebook are able to alter how an individual behaves, we're reaching a critical moment in how we want to coexist with computational systems. If we expose ourselves to technology that is programmed to be sexist, misogynistic, homophobic, racist, and so forth, we shouldn't be surprised to see things like *The Wall Street Journal*'s "Blue Feed, Red Feed," which shows you what you will see if Facebook tags you as liberal or conservative. What you see is what you might easily become today. Try scrolling through the feed of a stranger sitting right next to you, and you'll see that their online reality can be a lot different from yours. That's because today we get the news we "want"—we get to be hedonically stimulated to confirm what we think is the truth, thus validating how smart we are: I get the Daily Me, and you get the Daily You. And then, when occasionally exposed to an "opposing point of view," we have no other recourse but to assume the other side's ignorance compared with our own superior views. Meanwhile, all along the machines are feeding this to us because we programmed them to do so. They're watching how we react positively or negatively to what we're fed, and in turn learn our individual extremes for good and bad.

But it's not too late to make computational machines that have a broader understanding of the human experience. We can easily start the process by first understanding ourselves better—this journey is under way for me right now as I dig deeply into the world of "inclusive design." This approach to leveraging diversity is the key to making better products. I'm embarrassed to say that at first I didn't fully understand why the area interested me so much, but in hindsight I realize it's because I could smell something. The rise of computational design, and the incredible business value it has brought with it, was also creating imbalances in ways that were not immediately apparent to the people and companies at the center of it all. The fact that AI *à la levure* was odorless really bothered me.

Fortunately there's change afoot, led by inclusive design expert Kat Holmes, with ideas that began when she was at Microsoft that are now spreading across the world via her 2018 book, *Mismatch: How Inclusion Shapes Design*. I first encountered Holmes's work a year prior to *Mismatch* and featured it in the "Design in Tech Report" back when it was just starting to take off. Now leading user experience design at Google, Holmes is poised to reshape the cloud in ways that I'm excited to see come true. Holmes's three design principles for addressing imbalance are simple enough to put into practice, and yet deep enough to spend a lifetime trying to master. They are:

1. "Recognize exclusion." Make a conscious effort to notice when someone or a group of people is being excluded. You'll need to consciously step into uncomfortable situations when doing so, but it's an easy task when considering how those who are being excluded already felt uncomfortable in the first place.

2. "Learn from human diversity." Go thick, and go into neighborhoods and cultures that are unlike your own. That means you need to leave the safety and comfort of your home or workplace and place yourself in danger or discomfort—which is a hard sell at first, but your return on investment will be high.

3. "Solve for one, extend to many." Construct solutions that break your biases and help you find new markets. Innovation is what achieves growth, and innovating is about bringing new perspectives to existing problems, it gets even better when entirely new problems get introduced

that wouldn't have been obvious without a different point of view.

Kat Holmes's framework helps to disrupt our natural biases to exclude with the positive intent to concentrate, focus, and deepen the solutions that we want to design for ourselves. Anything that we feel comfortable with is going to be laden with biases, and because computation has been built by techies, we can expect it to be laden with techie biases. Computation isn't the only medium infested with biases. Just before digital cameras, there was chemical photography—which was tuned for lighter skin tones. Or think about this: if you ask a Temple of Design acolyte to name ten masters of the Bauhaus school, they will surely name ten men—even though the Bauhaus was half men and half women. Or look to the number of movies directed by women, or the number of Fortune 500 CEOs who are women or decidedly "unpale." What's different about computation? You know it—it's incomplete. We can reshape it. We can improve it. We just need to start immediately.

4. Open source is a computational means to engineer equity.

Kat Holmes often points out the origin of the word "exclude": derived from Latin *excludere*, where *ex-* means "out" and *claudere* means "to shut." In the people world that translates to literally shutting out a group of people from a special club that lets everyone else in. It's hard to spin exclusion in a more positive way because it is unfair by its very nature—which you know if you've ever felt excluded or "shut out" yourself. But exclusion makes complete sense when considered from the point of view of business, because to have an "unfair advantage" is considered to be a winning weapon when competition is fierce. Having something that your competition does not incentivizes you to shut them out and adopt what's called in the computing world a "closed" approach.

It's a common practice in industry to make closed systems, because when successful they provide the invaluable ability to exercise full control. When launching the first Mac computer in 1984, Apple famously went with a closed computing system that wasn't easy to extend like the competing Wintel PC standard at the time. As a result, Apple was able to control the entire user experience in ways that no other computing brand could achieve. This strategy of a closed system approach played out again later with Apple's launch of the iPhone, and the rest is history. Meanwhile, there was an emerging mobile OS project called Android that chose the unusual path of making all of its computer code ("source code") openly available. Today, there are more devices powered by Android than Apple's operating systems. As of 2019, Apple's approach has started to weaken, and the company is being forced to participate outside its own closed universe. Are its unfair advantages eroding?

"Open source" is the official term for computer code that is open and accessible to anyone to modify for their own purposes. It's the opposite of "shutting out" others—instead, it's about including anyone and everyone. A few well-known open source projects include the operating system Linux (on which Android is built), Firefox (a popular web browser), WordPress (the website management system for over a third of web traffic), and PHP (the popular computer language that powers WordPress). The term "open source software" was coined by Christine Peterson in 1998 as a way to better embody the community values that are inherent in it, as opposed to the prevailing term of the time, "free software," which connotes lesser quality. Having had the opportunity to witness the WordPress community firsthand through my work at Automattic, I don't recall ever encountering a more welcoming, inclusive, and world-spanning group of people in my entire life. And over time, I came to realize that PHP in the WordPress universe stood for "People Helping People," given the way each local community welcomes anyone who wants to learn computation, with no strings or costs attached to getting involved as a contributor. In open source, the software is the community and not just the code.

In contrast, "closed source software" governs the majority of apps and services that you use every day. You'll never be able to examine what the programming code is actually doing, and if you want it to work differently it's impossible for you to make changes to the software. This includes your Facebook apps and most of everything running on your phone or desktop computer or online. Now, even if you could access the source code of all your apps, that doesn't mean you would automatically be able to understand what's there. The same goes for a complicated open source system like WordPress. But it's a fact that your Facebook app shuts you out of the opportunity to look under the hood, whereas WordPress is fully inclusive at the source code level if you ever want to change any aspect of how it works for you.

Another way to think about the difference between closed

source and open source is to consider the distinction between "co-operation" and "collaboration." Cooperation is about working with another party at arm's length, whereas collaboration is about having arms hugged around each other. The advantage of collaboration over cooperation is that mutual benefits result from working together, with all parties making compromises of varying degrees. In the absence of the ability to collaborate, the only recourse for governments to rein in the Temple of Tech today is to attempt to regulate it. Interestingly, if all software by Temple of Techsters were fully open source, then there would be no need for governments to take their current course of actions against them. Why? Because the source code would be inspectable for violations of issues that we are all concerned about today, like knowing what they're doing with all the data they are gathering about us. It's harder to do evil when there aren't opaque walls shutting everyone else out. An open systems approach is an alternative to government regulation, and so I expect we'll see more of this approach when politicians who can speak machine, like you, get elected. Maybe it's time for you to run for office? I hereby open source an OPEN campaign slogan for you, with a little recursive twist: "OPEN Promotes Equity, Naturally."

There is a downside to open source: there are no secrets to be kept anywhere. In a world where everyone seeks to collaborate with each other and inflict no harm, then full transparency can mean "sharing is caring" and lasting harmony. However, there will always be a few bad actors who are always looking for ways to manipulate a situation in their favor for reasons that can only be explained as human nature. So open source is not always the way to go. For example, you would never want to publish all the source code for your personal electronic banking system that can easily access all of your finances. An open source approach might be commendable if such an act of sharing let others make a similar system for themselves, but you can bet that all your money would soon vanish if your source code included sensitive information, like bank account numbers and passwords. Or if Facebook's algorithms were all open sourced, then an entity with malicious intent could rewrite the timeline codes and easily manipulate your timeline. And of course, there will always be competitive advantages to justify why a business would want to keep its code private: to keep its unfair advantages over its rivals.

Nonetheless, businesses are recognizing the value of open source. Microsoft surprised the world with its acquisition of GitHub, the world's largest community of open source software development. To grasp the magnitude of that acquisition, just ask your programmer friends about it—some may not even know that Microsoft owns GitHub, because Microsoft has chosen not to alter or rebrand how it currently operates. Besides Android, another example at Google is the Chrome web browser, which runs on an open source engine. Now anybody can make a web browser on top of it—and even Microsoft has announced it will be switching over to Chrome's engine. Relatedly, Apple's web browser Safari is an example of a hybrid open/closed system: it shares its lineage with the same open source engine as Google, but the rest of its code is not accessible to the public. When using open source in your product, be sure to check the licensing rules—some licenses let you use the code without any restrictions, while others require you to openly share the rest of your code if you use theirs. The former is often referred to as an "MIT license," which gives you a lot of freedom; the latter is a GNU "GPL license," which is more about giving others a lot of freedom.

Let's not forget about the other kind of programming that has less to do with shareable computer codes—I'm talking specifically about AI *à la levure*. Newer machine intelligence systems aren't composed of readable computer codes but are instead packaged as opaque black boxes of numbers and data with no clear logical flow. It's long been a concern that these methods are so complex that we don't really know how they work—they're not legible by human beings because they're essentially piles of raw numbers. The severely closed nature of these systems—which have biases based in the data that they are fed to be trained—has set off alarm bells about the need to address their inherent opacity. New work is now emerging on computational ways to inspect these opaque AIs to behave more like "gray boxes" that might give us more insight into how they work. And if we can't figure out how they work, there are also efforts under way for AIs to start asking *why* they're being told to do something, so they might build the equivalent of a conscience. We should expect and demand more efforts around both understanding AIs and teaching ethics to AIs while channeling the machines' never-ending diligence to loop forever until they succeed.

Frankly, it's easy to be terrified of AI as it's becoming an increasingly common topic in popular media. We're not far away from hearing about a cleaning robot that refuses to listen to you, or an app-enabled pacemaker that extorts you, or a cybercrime cartel that has replaced your entire online presence with a bot you can't control—none of these has happened yet, but all are entirely possible with existing technology. If and when such calamities do occur, just remember that computation right now is just one of two things: readable source code or black boxes of numbers. It's *au levaine* or *à la levure*. Both are made by human beings as IF-THEN logic or data-powered black boxes, and they're either openly shared or hidden behind closed doors.

When they're open technologies, we have the opportunity to share, collaborate, and learn together. And when we share similar values with the cocreators of the open source code, we are less fearful of the technology. If you join an open source community, you'll feel the responsibility to do the right thing by everyone in it. Most of the communities don't require you to be an expert computer programmer, and even rudimentary "machine speakers" are more than welcome. In case you would like to get involved in one of the many open source communities to grow your machine-speaking abilities, I've compiled a list on howtospeakmachine .com. Why do you want to get involved? OPEN Promotes Equity, Naturally.

5. Mind the humans.

Shortly after four a.m. on December 6, 2015, I went out for a jog on El Camino Real in Palo Alto. It was like every early morning run for me—not too cold, not too hot. Dry. Safe. Along a route I knew well. On my mind was a six a.m. phone call I needed to make after my run. There were no cars out and about, but when I was making my way across a crosswalk on a six-lane road, the light began to turn to red, so I sped up. The other side of the street was dark, as it always was. I made it across the street before the light turned red, feeling slightly victorious, when my right foot caught the edge of the sidewalk.

And I tripped.

I landed on the sidewalk flat on my face, arm, and knee. My head was ringing. It was still quiet and dark, with nobody else around. I touched my face with my left hand. It felt wet and I quickly surmised that I was bleeding. My right hand couldn't move, and I soon realized that something had happened to my right elbow, as it couldn't go straight. My elbow felt a bit like little Lego pieces. I was scared.

A few cars passed by hastily. I was wearing random startup swag black. No phone with me. And no wallet or ID. I had my new Apple Watch to measure my steps... but it couldn't help me call for help as it was still the first version. I started to shiver in shock. I knew I needed to get back to my Airbnb—which was roughly ten California-size blocks away. There was nobody else on the streets. I also thought that nobody was likely to help me as I was dressed in a dark hoodie, bleeding from my face, and really... what commuter in a hurry would want to stop to help this creature fresh out of a horror movie? As I looked up into the dark sky, I felt oddly at ease because it struck me how insignificant I was—just another random organism on the surface of the planet, of no more significance than anyone else. It was a wonderfully humbling feeling. I felt peacefully in pain.

My MIT-trained engineering mind kicked in all of a sudden. Inexplicably, I imagined myself to be a Mars autonomous rover that had a few broken parts and needed to get back to base for repair. Engineers know well how those units are equipped with many redundant systems in case any systems fail. And so I imagined that there must be some way I could get back to my Airbnb too. This mental image of "becoming a machine" helped me completely forget about the pain. The adrenalin probably helped too.

I soon realized that I couldn't go a few steps without passing out, so I simply got up, took a few steps, and then I would lie on the ground. I made continual progress in this manner. When I started to turn off the broad concrete fairway of El Camino Real, the soft, relaxing feeling of placing my face against the grass on a few neighbors' lawns kept me motivated to continue forward.

Fortunately, I got back to the Airbnb, to my phone, located the nearest hospital with a call to my assistant, tried to clean off some of the blood, and called an Uber. When I got to the ER around five thirty a.m., I was handed a clipboard and pencil. My good arm was broken and I am a righty, but I quickly started to adapt as a lefty to my best abilities and filled out the form with the penpersonship of a second grader.

I anxiously waited in one of the inpatient rooms for about an hour, wanting nothing more than to see a doctor. Finally, a doctorlike person walked in and looked at my torn face where my front teeth had broken through my upper lip. "You look terrible!" he said. I figured he couldn't be a doctor with such a bedside manner, but then again I couldn't be sure. He then said, "Can you move your neck?" I did so. With a serious look, he then exclaimed, "You're lucky!" *And at that moment, I felt immediately relieved.* I agreed, "I'm lucky!" I thought to myself how bad it would have been if I couldn't move my neck—I would have been stuck on the sidewalk and unable to walk. I would have been an entirely broken machine. I felt so grateful. He then stitched up my face.

Half an hour later a nurse came in, looked at me, and asked, "What happened?"

I replied, "I was jogging and tripped."

In a serious, scolding tone, he said, "Exercise isn't good for you. Don't you know that?"

I tried to smile but the anesthesia from the stitches didn't let me.

He then said, "Were you wearing a fluorescent vest or a light?"

"No," I said.

"You could have been hit by a car!"

Another flash of joy came as I thought, *Yeah. I really could have been hit by a car.* For two years I had been jogging in the morning darkness wearing all black—which seemed brainless in hindsight. I could easily picture myself getting hit by one of those popular, speedy (and silent) electric cars in Silicon Valley due to my carelessness. So I felt even more relieved and happy!

The next day, as I was being wheeled into the operating room and with the anesthesia just starting to kick in, my epiphany came. Just as I started to see the twilight pink sending me off to sleep, I felt a jolt of realization that *the computational era needed to address the imbalance between technology and humans.*

Although my path to recovery involved a great deal of technology, it was my religious moment of awakening that drove me to pay careful attention to all the people in my direct surroundings. So instead of marveling at the latest technology that was repairing me, I observed the many compassionate human beings working alongside the machines. The doctors, nurses, technicians, receptionists, my Airbnb hosts Betty and Benny who took care of me after my operation, cleaners, food service workers, the flight attendant who lifted my roll-about bag into the overhead bin when I realized I couldn't do it myself, and a whole host of nonmachines who eventually got me back to work. I also was conscious of the people who indirectly impacted me, like all the engineering, design, and product folks who I'd never had a chance to meet, and all the invisible teams out there who had shipped the machines (and parts) that helped to repair me. My recovery took ten months in fits and starts, and it definitely did not transpire at Moorean speed. But I would do it all over again because the journey was absolutely worth it.

Recognizing one's own humanity while recognizing the humanity of others is the kind of gift that technology cannot give to you. I wouldn't wish sickness on anybody, and yet whenever someone asks me why I care so much about inclusion, I suggest that they break a bone in their body. Their response is usually a polite "No, thanks." And yet I'll continue to insist that they do, because that was what enabled me to fully acknowledge the privilege I have been given and earned. During my recovery I was often filled with an overwhelming sense of being lucky to have been born into a family with parents who sacrificed everything they had so I could go to a special place like MIT to learn everything I'd ever need to know about machines. They never had access to the kind of health care that I enjoyed in my special position in society—and finding this sense of gratitude along with the accountability it brings to the humans around us, and who have enabled us, is what has landed and planted itself firmly in the foreground for me.

I guess that's the one last thought I'd like to leave with you as you head out with your new language skills, now that you know how to speak machine too. As a fellow computational thinker, never forget: mind the humans. We are the ones who brought the computational era into existence. And we're grappling with what that means for computational products and services today that ship incomplete and instrumented. Now more than ever we need to think and work inclusively in order to directly address the imbalances that will be automated if we don't consciously create new paths.

It all comes back to the difference between cooperation and collaboration:

COOPERATION	COLLABORATION
= working together independently	= working together dependently

Working cooperatively is easier than working collaboratively, because to cooperate you don't really have to understand the other party deeply. For most of the history of the computer's evolution, the majority world of everyday humans has been learning to cooperate and cope with machines that keep changing for some unknown reason. Meanwhile, it's the comparatively smaller group of computational thinkers in Silicon Valley and the like that has instead been working collaboratively with the omniscience lurking in the cloud and knowingly at Moorean speeds. And we're trusting them all to collaborate well on behalf of humanity. Maybe until you had read this book you couldn't easily collaborate with computers, or with their human collaborators directly. That's understandable because you hadn't yet taken the time to visit their invisible universe to learn their history, customs, and norms. You couldn't speak the language of the machine at all. Now you do. If just a "bit."

Our machines run loops. Our machines operate at infinitely large and infinitesimally small scales. Our machines are becoming alive. Our machines are incomplete and imperfect, like us. Our machines are increasingly instrumented and know what we're up to.

Our machines are automating imbalance all over the world on our watch. Mind the machines. Mind the humans. Let's go.

Acknowledgments

This book is dedicated to my parents, Elinor "Yumi" and Yoji. If they hadn't somehow saved enough money to get me a computer, I would never have had the chance to understand the computer the way I do today.

My editor, Niki Papadopoulos, gave me the license and encouragement to write this book—for which I'm deeply grateful. Niki gave me a psychologically safe space to fully explore the ideas that you read here. Rafe Sagalyn's visionary hand was constant, while Laura Parker was my writing coach along the way. Rebecca Shoenthal, together with the Portfolio folks, enthusiastically carried this entire project to the finish line. I'm thankful to the entire team of people that brought this book together in the form in which you have before you now. I owe an especially big thank you to the technical reviewers of this book: Kevin Bethune, Tracy Chou, Martin Eriksson, Alexis Lloyd, Rochelle King, and Theresa Austin. They all helped me to find many bugs I had littered throughout the pages of the manuscript.

There are folks whose ideas and encouragement have influenced my thinking on the road to understanding computation as a humanist. They were all people I encountered during my MIT days as heroes, and, in some cases, I was lucky to have them as my mentors. I don't know about you, but I find that the problem with getting older is that people you once looked up to tend to disappear, like Muriel Cooper, Paul Rand, Ikko Tanaka, William J. Mitchell, Robert Silbey, Red

Burns, Charles M. Vest, Whitman Richards, Mits Kataoka, and, most recently, my 1980s MIT AI professor and former colleague, Patrick Henry Winston. I feel fortunate to have encountered their ways of thinking while they were still alive, because if you know the difference between a fresh loaf of bread versus a really stale one, it's hard to settle for anything less than fresh from the oven.

After getting the computation bug (and bugs) out of my system, my interests are turning toward another area that's fascinated me: organizations and how they transform. Or don't. Given that Moore's law continues to have a doubling impact on our world, it's only prudent that organizations start changing. Fast. And, preferably, exponentially fast. My thoughts on this matter are influenced by the work of Regina Dugan, Hugo Sarrazin, Ivy Ross, Jason Fried, Kim Scott, Scott Belsky, Beth Comstock, Becky Bermont, Cat Noone, Hajj Flemings, Maria Giudice, David Granger, Marina Mihalakis, Betty and Benny Xian, Georgia Frances King, Katharine Schwab ... as I type this list I realize it will be incomplete, so it's only natural that I should iterate on it over on the book's website at howtospeakmachine.com.

Lastly, I wish to acknowledge Ben Ichinose (1929–2019). As a complete stranger, Ben cold-called me shortly after the publication of *The Laws of Simplicity* (MIT, 2006). He wanted to show me his backyard where, as a retired orthodontist, he spent years hand-constructing a traditional Japanese garden to represent his principles of simplicity. After his many calls and my many refusals, I finally went to see his backyard. Wearing his signature aviator sunglasses, Ben opened a 1966 bottle of wine to toast me while he broke a fresh loaf of bread. If there was anyone I've met in my life who embodied the oddly enchanting aspects of computation, it was Ben.

Notes

INTRODUCTION

xiii *"I have always been interested":* John Maeda, "Why Simplicity?," *Maeda's Simplicity* (blog), MIT Media Lab, December 17, 2004, web.archive.org/web /20041230214607/http://weblogs.media.mit.edu:80/SIMPLICITY/ar chives/000045.html.

xiv *I knew that the controversy:* John Maeda, "Fast Company on the Design in Tech Report," *Design in Tech Report* (blog), March 24, 2019, designintech .report/2019/03/24/fast-company-on-the-design-in-tech-report-%F0% 9F%90%B8-edition.

xvi *the late musician David Bowie:* BBC Newsnight, "David Bowie Speaks to Jeremy Paxman on BBC Newsnight (1999)," YouTube, youtube.com/watch?v= FiK7s_0tGsg&t=665s.

xviii *the eeriness of tech:* James Poniewozik, "'Black Mirror' Finds Terror, and Soul, in the Machine," *The New York Times*, October 20, 2016, nytimes.com /2016/10/21/arts/television/review-black-mirror-finds-terror-and-soul -in-the-machine.html.

CHAPTER 1

16 *a definition that goes back to 1613:* John Maeda, "First Use of the Word 'Computer,'" *How to Speak Machine* (blog), February 24, 2019, howtospeakma chine.com/2019/02/24/first-use-of-the-word-computer.

16 *the 1895 Century Dictionary:* wordnik.com/words/computer.

18 *the ENIAC (Electronic Numerical Integrator and Computer):* seas.upenn.edu /about/history-heritage/eniac/.

18 *the women computers of ENIAC:* "ENIAC Programmers Project," First Byte Productions, LLC, 2019, eniacprogrammers.org/eniac-programmers-project/.

24 *called his endeavor the GNU Project:* gnu.org/gnu/gnu-history.en.html.

26 *expressed succinctly by Michael Corballis:* Michael Corballis, *The Recursive Mind: The Origins of Human Language, Thought, and Civilization* (Princeton, NJ: Princeton University Press, 2011), 1.

CHAPTER 2

48 *designers Ray and Charles Eames:* "Powers of Ten and the Relative Size of Things in the Universe," Eames Office, LLC, 2019, eamesoffice.com/the -work/powers-of-ten/.

52 *area covered by this snowflake:* Evelyn Lamb, "A Few of My Favorite Spaces: The Koch Snowflake," *Scientific American*, November 30, 2017, blogs .scientificamerican.com/roots-of-unity/a-few-of-my-favorite-spaces-the -koch-snowflake/.

55 *"The computer programmer is a creator":* John Maeda, "Joseph Weizenbaum: Human Technologist," *John Maeda's Blog* (blog), February 11, 2019, maeda .pm/2019/02/11/joseph-weizenbaum-humanist-technologist.

60 *a live image feed of a coffeepot:* Quentin Stafford-Fraser, "The Life and Times of the First Web Cam," *Communications of the ACM* 44, no. 7 (July 2001): 25–26, cl.cam.ac.uk/coffee/qsf/cacm200107.html.

60 *In 1994, there were 2,738 websites:* internetlivestats.com/total-number-of -websites/.

64 *Netflix would also want to shop:* Kevin McLaughlin, "Netflix, Long an AWS Customer, Tests Waters on Google Cloud," *The Information*, April 17, 2018, theinformation.com/articles/netflix-long-an-aws-customer-tests-waters -on-google-cloud.

CHAPTER 3

66 *Scientist Valentino Braitenberg:* mitpress.mit.edu/books/vehicles.

67 *Braitenberg went on to design:* mitpress.mit.edu/books/vehicles.

67 *"However far modern science and techniques":* Lewis Mumford, *Technics and Civilization* (New York: Harcourt, Brace and Company, 1934), 435.

68 *It's the disembodied robot:* John Maeda, "Abstract," *How to Speak Machine* (blog), April 13, 2019, howtospeakmachine.com/2019/04/13/us-patent-4058672-and -voice-prompting-1976/.

69 *And if it's got more stamina:* Louis Liebenberg, "Persistence Hunting by Modern Hunter-Gatherers," *Current Anthropology*, 47.6 (2006): 1017–1026.

69 *among those capable of telling the difference:* James Vincent, "Watch Jordan Peele Use AI to Make Barack Obama Deliver a PSA About Fake News," *The Verge*, April 17, 2018, theverge.com/tldr/2018/4/17/17247334/ai-fake-news -video-barack-obama-jordan-peele-buzzfeed; Karen Hao, "Google's AI Can Now Translate Your Speech While Keeping Your Voice," *MIT Technology Review*, May 20, 2019, technologyreview.com/s/613559/google-ai-language-translation.

70 *"person-centered therapy":* psychologytoday.com/us/therapy-types/person -centered-therapy.

73 *Weizenbaum spent his years:* "Professor Joseph Weizenbaum: Creator of the 'Eliza' program," *The Independent*, March 18, 2008, independent.co.uk

/news/obituaries/professor-joseph-weizenbaum-creator-of-the-eliza -program-797162.html.

74 *"artificial neural networks":* F. Rosenblatt, "The Perceptron: A Probabilistic Model for Information Storage and Organization in the Brain," *Psychological Review* 65, no. 6 (1958): 386–408.

78 *In Weizenbaum's era of AI:* Istvan S. N. Berkeley, "A Revisionist History of Connectionism," 1997, userweb.ucs.louisiana.edu/~isb9112/dept/phil341 /histconn.html.

78 *"'automation on steroids.'":* Andrew Ng (@AndrewYNg), 2017, "If you're trying to understand AI's near-term impact, don't think 'sentience.' Instead think 'automation on steroids,'" Twitter, May 1, 2017, twitter.com/andrewyng/sta tus/859106360662806529?lang=en.

78 *prior to 2012:* Benedict Evans, "Voice and the Uncanny Valley of AI," *Benedict Evans* (blog), March 9, 2017, ben-evans.com/benedictevans/2017/2/22/voice -and-the-uncanny-valley-of-ai.

80 *the majority of plants on our planet:* John H. Lienhard, "No. 1243: Five-Fold Symmetry," *The Engines of Our Ingenuity* (blog), uh.edu/engines/epi1243 .htm.

82 *curate my own little desktop nature lab:* Zhenqiang Gong, Nichilos J. Matzke, Bard Ermentrout, Dawn Song, Jann E. Vendetti, Montgomery Slatkin, George Oster, "Evolution of patterns on Conus shells," Proceedings of the National Academy of Sciences, January 2012, 109 (5) E234-E241; DOI: 10.1073 /pnas.1119859109.

88 *live inside the majesty:* Sarah Zielinski, "The Secrets of Sherlock's Mind Palace," Smithsonian.com, February 4, 2014, smithsonianmag.com/arts-culture /secrets-sherlocks-mind-palace-180949567.

88 *"the freedom to be lords":* David Foster Wallace, "Plain Old Untrendy Troubles and Emotions," *The Guardian,* September 19, 2008, theguardian.com /books/2008/sep/20/fiction.

90 *"A single twig breaks":* "Intermediate Level Lessons: Bundle of Twigs," *War of 1812,* PBS, pbs.org/wned/war-of-1812/classroom/intermediate/bundle -twigs.

90 *In the mathematical world:* Martin Gardner, "Mathematical Games: The Fantastic Combinations of John Conway's New Solitaire 'Game Life,'" *Scientific American* 223 (October 1970): 120–23. ibiblio.org/lifepat terns/october1970 .html.

90 *Conway's "Life" is immediately misleading:* corp.hasbro.com/news-releases /news-release-details/game-life-celebrates-50-years.

93 *the human body comprises trillions:* "NIH Human Microbiome Project Defines Normal Bacterial Makeup of the Body," National Institutes of Health, June 13, 2012, nih.gov/news-events/news-releases/nih-human-microbiome -project-defines-normal-bacterial-makeup-body.

94 *So putting chess in the spotlight:* Nathan Ensmenger, "Is Chess the Drosophila of Artificial Intelligence? A Social History of an Algorithm," *Social Studies of Science* 42(1) (2011): 5–30, pdfs.semanticscholar.org/c9e7/3fc7ec81458057e6 f96de1cba095e84a05c4.pdf.

94 *the promise of AI:* National Research Council, *Funding a Revolution: Government Support for Computing Research* (Washington, DC: The National Academies Press, 1999), 143. doi.org/10.17226/6323.

94 *But the high expectations set:* Eleanor Cummins, "Another AI Winter Could Usher in a Dark Period for Artificial Intelligence," *Popular Science*, August 29, 2018, popsci.com/ai-winter-artificial-intelligence/.

94 *the return of government and industrial funding:* "DARPA Announces $2 Billion Campaign to Develop Next Wave of AI Technologies," Defense Advanced Research Projects Agency, September 7, 2018, darpa.mil/news-events/2018 -09-07.

96 *science fiction writer and computer scientist:* Vernor Vinge, "The Coming Technological Singularity: How to Survive in the Post-Human Era," 1993, edoras.sdsu.edu/~vinge/misc/singularity.html.

96 *Ray Kurzweil went one step further:* Ray Kurzweil, "The Law of Accelerating Returns," Kurzweil Accelerating Intelligence/Essays, March 7, 2001, kurz weilai.net/the-law-of-accelerating-returns.

97 *for AI pioneers like the late MIT professor:* "The Singularity—2045: The Year Man Becomes Immortal," *Time*, 2019, content.time.com/time/interactive /0,31813,2048601,00.html.

97 *Singularity is just another big idea:* "Marvin Minsky: Why I Prefer Science Fiction to General Literature," *Web of Stories*, YouTube, January 4, 2017, youtube.com/watch?v=c8Af6Y6HBCE.

97 *a mouse brain was reported to have been simulated:* Moheb Costandi, "Fragment of Rat Brain Simulated in Supercomputer," *Nature*, October 8, 2015, nature.com/news/fragment-of-rat-brain-simulated-in-supercomputer-1 .18536.

98 *experience more computational surrogates:* Philip van Allen, "2011: Object Animism," *Media Design Practices* (blog), mediadesignpractices.net/lab/re searchproject/object-animism-philip-van-allen-faculty-researcher.

CHAPTER 4

102 *you'll hear the word "agile":* Kent Beck et al., "Manifesto for Agile Software Development," 2001, agilemanifesto.org.

102 *You'll hear "lean":* Jeff Gothelf, "Tag: Lean Startup," *Jeff Gothelf* (blog), December 15, 2016, jeffgothelf.com/blog/tag/lean-startup.

102 *And "scrum"?*: Emam Hossain, et al., "Using Scrum in Global Software Development: A Systematic Literature Review," 2009 Fourth IEEE International Conference on Global Software Engineering, ieeexplore.ieee.org /abstract/document/5196931.

105 *form of "planned obsolescence."*: John Maeda, "Planned Obsolescence," *John Maeda's Blog* (blog), December 26, 2018, maeda.pm/2018/12/26/planned -obsolescence.

107 *But knowing your ABCs:* Hans M. Wingler, *Bauhaus* (Cambridge, MA: MIT Press, 1969), mitpress.mit.edu/books/bauhaus.

108 *proverbial ten thousand hours:* K. Anders Ericsson, Ralf Th. Krampe, and Clemens Tesch-Romer, "The Role of Deliberate Practice in the Acquisition of Expert Performance," *Psychological Review* 100, no. 3 (1993): 363–406, proj ects.ict.usc.edu/itw/gel/EricssonDeliberatePracticePR93.pdf.

110 *"The new human should":* John Maeda, "Tech x Business x Design," *Design in Tech Report* (blog), March 10, 2019, designintech.report/2019/03/10/%F0% 9F%93%B1design-in-tech-report-2019-section-1-tbd-tech-x-business -x-design.

113 *my new approach is:* John Maeda, "Why I Use SMS to Integrate Q&A into My Talks Instead of After the Talk," *John Maeda's Blog* (blog), March 21, 2018, maeda.pm/2018/03/21/why-i-use-sms-to-integrate-qa-into-my-talks-instead -of-after-the-talk.

113 *keeping the celebrated Japanese essay:* Maria Popova, "In Praise of Shadows: Ancient Japanese Aesthetics and Why Every Technology Is a Technology of Thought," *Brain Pickings*, May 25, 2015, brainpickings.org/2015/05/28/in -praise-of-shadows-tanizaki/.

114 *curator Paola Antonelli dared to collect:* Esther Zuckerman, "Video Games: Art-Tested, MoMA-Approved, *The Atlantic*, November 29, 2012, theatlantic.com /entertainment/archive/2012/11/video-games-art-tested-moma-approved /321022.

114 *an open digital archiving solution:* Dragan Espenschied, "Rhizome Releases First Public Version of Webrecorder," *Rhizome* (blog), August 9, 2016, rhi zome.org/editorial/2016/aug/09/rhizome-releases-first-public-version -of-webrecorder.

116 *embodied by contrarian fashion icons:* Catherine Clifford, "Iris Apfel: 10 life lessons from a 96-year-old who is probably cooler than you," CNBC, March 29, 2018, cnbc.com/2018/03/29/10-life-lessons-from-96-year-old-iris-apfel.html.

117 *venture capital funding:* John Maeda, "Start-ups vs End-ups (2013)," *John Maeda's Blog* (blog), June 19, 2016, maeda.pm/2016/06/19/start-ups-vs-end -ups-2013.

120 *"The more useful your software is":* "Inertia in the interface," June 27, 2019, Jessica Kerr, blog.jessitron.com/2019/06/27/inertia-in-the-interface/.

122 *compounded losses that can cause:* Dana Olsen, "The State of US Venture Capital in 15 Charts," *PitchBook*, October 29, 2018, pitchbook.com/news/articles/the-state-of-us-venture-capital-activity-in-15-charts.

122 *"time value of shipping,":* Brandon Chu, "Product Management Mental Models for Everyone," *The Black Box of Product Management*, August 19, 2018, blackboxofpm.com/product-management-mental-models-for-everyone-31e7828cb50b.

123 *"Speed and thoughtfulness need to coexist":* John Maeda, "Speed x Thoughtfulness," *How to Speak Machine* (blog), June 2, 2019, howtospeakmachine.com/2019/06/02/speed-x-thoughtfulness.

123 *vaguely similar to a vastly more favorable phrase:* "Fire and Forget," Wordnik, wordnik.com/words/fire%20and%20forget.

126 *The minimum viable product:* "MVP: Minimum Viable Product," SyncDev, syncdev.com/minimum-viable-product/.

126 *the minimal or "lean":* Taiicho Ohno, *Toyota Production System: Beyond Large-Scale Production* (Boca Raton, FL: CRC Press, 1988).

126 *Although the definition of "viable":* John Maeda, "Minimum Desirable Product," *John Maeda's Blog* (blog), April 21, 2016, maeda.pm/2019/04/21/minimum-desirable-product.

127 *I like to use the term "MVLP":* John Maeda, "Minimum Viable Lovable Product (MVLP)," Design.co, December 27, 2018, design.co/2018/12/27/minimum-viable-lovable-product-mvlp.

129 *former Yahoo CEO Marissa Mayer:* O'Reilly, "Velocity 09: Marissa Mayer, In Search of . . . a Better, Faster, Stronger Web," YouTube, youtube.com/watch?v=WFsQvcdmLxc.

129 *users "really respond to speed,":* O'Reilly, "Velocity 09: Marissa Mayer, In Search of . . . a Better, Faster, Stronger Web," YouTube, youtube.com/watch?v=WFsQvcdmLxc.

129 *a material increase in traffic:* Dan Farber, "Google's Marissa Mayer: Speed Wins," *ZDNet*, November 9, 2006, zdnet.com/article/googles-marissa-mayer-speed-wins.

129 *even though most celebratory articles:* Joshua Topolsky, "Google Is Really Good at Design," *The Outline*, October 12, 2017, theoutline.com/post/2388/google-is-really-good-at-design?zd=1&zi=nzxsnl3x; Cliff Kuang, "How Google Finally Got Design," *Fast Company*, June 1, 2015, fastcompany.com/3046512/how-google-finally-got-design.

130 *a tasty burger at breathtaking speed:* David Brancaccio, "The True Origin Story Behind McDonald's," Marketplace, February 9, 2017, marketplace.org/2017/02/09/business/ray-kroc-mcdonalds-fast-food.

130 *their Speedee Service System:* "The McDonald's Story," McDonald's, corporate.mcdonalds.com/corpmcd/about-us/history.html.

130 *"Savings in time feel like simplicity."*: John Maeda, "Law 3/ TIME," The Laws of Simplicity, lawsofsimplicity.com/los/law-3-time.html.

131 *miniature version of a bigger cake:* Brandon Schauer, "Cupcakes: The Secret to Product Planning," Adaptive Path, February 10, 2011, web.archive.org /web/20150922090415/adaptivepath.org/ideas/cupcakes-the-secret -to-product-planning.

CHAPTER 5

134 *the snow depth from Mont Blanc:* Wilfrid J. Mayo-Wells, "The Origins of Space Telemetry," *Technology and Culture* 4, no. 4 (December 1963): 499, research gate.net/publication/269657057_The_Origins_of_Space_Telemetry.

135 *if it's instrumented:* "Good Design Is Good Business," McKinsey & Company, October 2015, mckinsey.com/business-functions/organization/our-insights/good -design-is-good-business.

137 *General Data Protection Regulation (GDPR):* Sarah Gordon and Aliya Ram, "Information Wars: How Europe Became the World's Data Police," *Financial Times*, May 20, 2018, ft.com/content/1aa9b0fa-5786-11e8-bdb7 -f6677d2e1ce8.

139 *a word my father often used:* "Omotenashi: The Reason Why Japanese Hospitality Is Different," *Michelin Guide*, April 4, 2017, guide.michelin.com/sg /features/omotenashi/news.

140 *a famous story that illustrates this best:* Yuko Daishoji, "The Three Cups of Tea," Miyoshi Tea Co., November 19, 2018, miyoshitea.com/new-blog/the -three-cups-of-tea-sankoncha.

140 *The tea server, Mitsunari Ishida:* The Editors of *Encyclopædia Britannica*, "Ishida Mitsunari," *Encyclopædia Britannica*, January 1, 2019, britannica .com/biography/Ishida-Mitsunari.

143 *this is currently fully legal:* "House Votes to Allow Internet Service Providers to Sell, Share Your Personal Information," *Consumer Reports*, May 4, 2018, consumerreports.org/consumerist/house-votes-to-allow-internet -service-providers-to-sell-share-your-personal-information.

144 *that this is really big:* "Patterns," Privacy Patterns, privacypatterns.org/pat terns; "An Evolving Collection of Design Patterns for Sharing Data," Data Permissions Catalogue, catalogue.projectsbyif.com; Holly Habstritt Gaal, "Ethical, by Design: How We Design with Your Privacy in Mind," *Duck-DuckGo* (blog), January 22, 2019, spreadprivacy.com/ethical-by-design.

144 *an innocent call to a local pizzeria:* "ACLU Links Pizza Delivery to Privacy Erosion in New Online Video," ACLU, July 26, 2004, aclu.org/news/aclu -links-pizza-delivery-privacy-erosion-new-online-video; ACLU, "Scary Pizza," YouTube, youtube.com/watch?v=33CIVjvYyEk.

145 *to receive their* omotenashi: "Guest Story: Expressed and Unexpressed Need," The Ritz-Carlton Leadership Center, June 24, 2015, ritzcarltonleadershipcenter.com/2015/06/expressed-and-unexpressed-needs.

147 *log a billion walking steps:* "1,000,000,000 Lifetime Steps . . . Is It Possible?," LetsRun.com, January 30, 2017, letsrun.com/forum/flat_read.php?thread= 8035415.

148 *"Be nice to nerds.":* "The Only Way to Learn to Fly Is to Fly: Regina Dugan at TED2012," *TED* (blog), February 29, 2012, blog.ted.com/the-only-way-to -learn-to-fly-is-to-fly-regina-dugan-at-ted2012.

148 *the role of data scientist:* Thomas H. Davenport and D. J. Patil, "Data Scientist: The Sexiest Job of the 21st Century," *Harvard Business Review,* October 2012, hbr.org/2012/10/data-scientist-the-sexiest-job-of-the-21st-century.

148 *"more enduring will be the need":* Davenport and Patil, "Data Scientist: The Sexiest Job of the 21st Century."

151 *"The best way to assess":* Erika Hall, *Just Enough Research* (New York: A Book Apart, 2013), 103.

151 *junk food is one of the few things:* Priya Fielding-Singh, "Why Do Poor Americans Eat So Unhealthfully? Because Junk Food Is the Only Indulgence They Can Afford," *Los Angeles Times,* February 7, 2018, latimes.com/opinion/op-ed /la-oe-singh-food-deserts-nutritional-disparities-20180207-story.html.

151 *Integrating quantitative and qualitative data:* John Maeda, "Quant Is the Scaffolding. Qual Is the Clay," *How to Speak Machine* (blog), June 2, 2019, howtospeakmachine.com/2019/06/02/quant-is-the-scaffolding-qual -is-the-clay.

154 *in 2014 Facebook shared:* Adam D. I. Kramer, Jamie E. Guillory, and Jeffrey T. Hancock, "Experimental Evidence of Massive-Scale Emotional Contagion Through Social Networks," *PNAS,* June 17, 2014, pnas.org/content/111/24 /8788.

154 *Rather than shipping a final website:* Dan Siroker, "How Obama Raised $60 Million by Running a Simple Experiment," *Optimizely* (blog), November 29, 2010, blog.optimizely.com/2010/11/29/how-obama-raised-60-million-by-running -a-simple-experiment.

154 *testing variations of an email:* Joshua Green, "The Science Behind Those Obama Campaign E-Mails," *Bloomberg Businessweek,* November 29, 2012, bloomberg.com/news/articles/2012-11-29/the-science-behind-those -obama-campaign-e-mails#r=auth-s.

155 *these kinds of "split tests":* Ron Kahavi, "Practical Guide to Controlled Experiments on the Web: Listen to Your Customers Not to the HiPPO," EXP Platform, August 2007, exp-platform.com/practical-guide.

156 *At the heart of the testing approach:* Julia Kirby and Thomas A. Stewart, "The Institutional Yes," *Harvard Business Review,* October 2007, hbr.org/2007 /10/the-institutional-yes.

156 *like Amazon, Airbnb:* Jan Overgoor, "Experiments at Airbnb," *Airbandb Engineering & Data Science* (blog), May 27, 2014, medium.com/airbnb -engineering/experiments-at-airbnb-e2db3abf39e7.

156 *Netflix, lies the simple:* Cara Harshman, "2 Controversial Site Redesigns That Should Inspire You to A/B Test," *Optimizely* (blog), August 14, 2014, blog .optimizely.com/2014/08/14/2-alexa-500-site-redesigns-that-should -inspire-you-to-ab-test.

157 *"Almost any questions can be answered":* Claude C. Hopkins, *Scientific Advertising*, CreateSpace Independent Publishing Platform (reprinted 2010 from original 1923 text), amazon.com/Scientific-Advertising-Claude-C-Hopkins /dp/1453821082/ref=sr_1_1?s=books&ie=UTF8&qid=1392230311&sr=1-1& keywords=scientific+advertising+by+claude+hopkins.

159 *but thankfully we have Wolfram Alpha:* Wolfram Alpha Computational Intelligence, wolframalpha.com.

160 *broken the speed of light:* John Maeda, "60mph Versus the Speed of Light." *How to Speak Machine* (blog), April 23, 2019, howtospeakmachine.com /2019/04/23/60mph-versus-the-speed-of-light.

160 *unethical sources of data collection:* John Maeda, "Talk Data to Me. Or, First Party Data vs Second Party Data vs Third Party Data," *How to Speak Machine* (blog), April 23, 2019, howtospeakmachine.com/2019/04/23/talk-data-to-me -or-first-party-data-vs-second-party-data-vs-third-party-data.

160 *location information that's either broadcast:* Elizabeth Dwoskin and Craig Timberg, "Google Now Knows When Its Users Go to the Store and Buy Stuff," *The Washington Post*, May 23, 2017, washingtonpost.com/news/the-switch/wp /2017/05/23/google-now-knows-when-you-are-at-a-cash-register-and - how- much-you- are- spending/? noredirect= on& utm_term=.aec 577034177; Zeynep Tufekci, "Think You're Discreet Online? Think Again," *The New York Times*, April 21, 2019, nytimes.com/2019/04/21/opinion/com putational-inference.html.

161 *thus dubbed the Centaur era:* Nicky Case, "How to Become a Centaur," *Journal of Design and Science*, February 6, 2018 (updated), jods.mitpress.mit .edu/pub/issue3-case.

162 *visit to any other shopping:* Jarno Koponen, "Get Ready for a New Era of Personalized Entertainment," *TechCrunch*, April 13, 2019, techcrunch.com/2019 /04/13/get-ready-for-a-new-era-of-personalized-entertainment.

163 *Media Lab cofounder Nicholas Negroponte:* Cass R. Sunstein, *Republic.com* (Princeton, NJ: Princeton University Press, 2007), 1.

163 *As elected leaders begin to speak machine:* Brian Resnick, "Yes, Artificial Intelligence Can Be Racist," *Vox*, January 24, 2019 (updated), vox.com/science -and-health/2019/1/23/18194717/alexandria-ocasio-cortez-ai-bias; Simone Stolzoff, "Meet Andrew Yang, a 2020 US Presidential Hopeful Running Against the Robots," *Quartz*, December 7, 2018, qz.com/1485345/meet

-andrew-yang-a-2020-us-presidential-hopeful-running-against-the
-robots.

CHAPTER 6

168 *"I learned two particularly important lessons"*: "A Study on the Status of Women Faculty in Science at MIT," *The MIT Faculty Newsletter* (special edition), March 1999, http://web.mit.edu/fnl/women/women.html.

169 *there was only 21 percent*: Rachel Gutman, "The Origins of Diversity Data in Tech," *The Atlantic*, February 3, 2018, theatlantic.com/technology/archive/2018/02/the-origins-of-diversity-data-in-tech/552155.

169 *the high-tech sector employed*: Alison DeNisco Rayome, "5 Eye-Opening Statistics About Minorities in Tech," *TechRepublic*, February 7, 2018, techrepublic.com/article/5-eye-opening-statistics-about-minorities-in-tech.

169 *a study by the Center for Employment Equity*: Donald Tomaskovic-Devey and JooHee Han, "Is Silicon Valley Tech Diversity Possible Now?," University of Massachusetts Amherst, June 2018, umass.edu/employmentequity/silicon-valley-tech-diversity-possible-now-0.

169 *why these folks leave the tech industry*: "The 2017 Tech Leavers Study," Kapor Center, April 27, 2017, kaporcenter.org/tech-leavers.

171 *when an internal hiring tool*: Jeffrey Dastin, "Amazon Scraps Secret AI Recruiting Tool That Showed Bias Against Women," Reuters, October 9, 2018, reuters.com/article/us-amazon-com-jobs-automation-insight/amazon-scraps-secret-ai-recruiting-tool-that-showed-bias-against-women-idUSKCN1MK08G.

172 *from an organizational perspective*: Lee Fleming, "Perfecting Cross-Pollination," *Harvard Business Review*, September 2004, hbr.org/2004/09/perfecting-cross-pollination.

172 *The result is everything*: Sara Wachter-Boettcher, *Technically Wrong: Sexist Apps, Biased Algorithms, and Other Threats of Toxic Tech* (New York: W. W. Norton & Company, 2018), 31.

172 *when a popular social media company released*: Davey Alba, "Clearly Snapchat Doesn't Get What's Wrong with Yellowface," *Wired*, August 10, 2016, wired.com/2016/08/snapchat-anime-filter-yellowface.

173 *actively working to diversify*: Courtney Seiter, "Our Latest Pay Analysis: Examining Buffer's Gender Pay Gap in 2019," Buffer, March 29, 2019, open.buffer.com/gender-pay-gap-2019.

173 *moving from the narrow-mindedness*: Stefanie K. Johnson, "What 11 CEOS Have Learned About Championing Diversity," *Harvard Business Review*, August 29, 2017, hbr.org/2017/08/what-11-ceos-have-learned-about-championing-diversity.

173 *For all of Google's diversity:* Jessica Guynn, "Google Gets Tough on Harassment After James Damore Firing Roils Staff," *USA Today,* June 27, 2018, usatoday .com/story/tech/news/2018/06/27/google-toughens-rule-internal -harassment-after-james-damore-firing-roils-staff/738483002.

173 *business leader Annie Jean-Baptiste:* Michelle Darrisaw, "Google's Annie Jean-Baptiste Talks Diversity and Creating Inclusive Products for Under served Communities," *Essence,* May 14, 2018, essence.com/lifestyle/money -career/annie-jean-baptiste-google-diversity-careers-interview.

173 *Google is examining everything:* Tulsee Doshi, "Introducing the Inclusive Images Competition," *Google AI* (blog), September 6, 2018, ai.googleblog.com /2018/09/introducing-inclusive-images-competition.html.

174 *Imagine rebooting tech education:* Kimberly Weisul, "Half of This College's STEM Graduates Are Women. Here's What It Did Differently," *Inc.,* May 31, 2017, https://www.inc.com/kimberly-weisul/how-harvey-mudd-college -achieved-gender-parity-computer-science-engineering-physics.html.

175 *how Advanced Placement computer science exams:* Ryan Suppe, "Female, Minority Students Took AP Computer Science in Record Numbers," *USA Today,* August 27, 2018, usatoday.com/story/tech/news/2018/08/27/female -minority-students-took-ap-computer-science-record-numbers/1079 699002.

177 *Cultural anthropologist Clifford Geertz:* Clifford Geertz, *The Interpretation of Cultures* (New York: Basic Books, 1973).

178 *when presented with quantitative data:* Tricia Wang, "Why Big Data Needs Thick Data," *Ethnography Matters* (blog), January 20, 2016, medium.com /ethnography-matters/why-big-data-needs-thick-data-b4b3e75e3d7.

182 *"having the rigor to 'unlearn'":* James Gadsby Peet, "Great Leaders Know When to Unlearn the Past," *Mind the Product* (blog), November 20, 2018, mind theproduct.com/2018/11/great-leaders-know-when-to-unlearn-the-past.

182 *gather as many observations from people:* Rochelle King, Elizabeth Churchill, and Caitlin Tan, *Designing with Data* (Sebastopol, CA: O'Reilly Media, 2017).

182 *When the cofounder of Google talks:* James Vincent, "Google's Sergey Brin Warns of the Threat from AI in Today's 'Technology Renaissance,'" *The Verge,* April 28, 2018, theverge.com/2018/4/28/17295064/google-ai-threat -sergey-brin-founders-letter-technology-renaissance.

183 *"Identity politics is market research":* MIT Media Lab (@medialab), "Identity politics is market research, which has been driving American politics since the 1930s. What platforms like Facebook have done is automate it.—Jill Lepore #MLTalks," Twitter, April 24, 2018, twitter.com/medialab/status /988866580498051072.

183 *crime prediction algorithms that tell us:* Daniel Cossins, "Discriminating Algorithms: 5 Times AI Showed Prejudice," *New Scientist,* April 12, 2018, newsci

entist.com/article/2166207-discriminating-algorithms-5-times-ai -showed-prejudice.

184 *crime-sentencing algorithms like COMPAS:* Jeff Larson, Surya Mattu, Lauren Kirchner, and Julia Angwin, "How We Analyzed the COMPAS Recidivism Algorithm," ProPublica, May 23, 2016, propublica.org/article/how-we -analyzed-the-compas-recidivism-algorithm.

184 *When asked about AI and its ramifications: The Fix,* Season 1, Episode 3, "Let's Fix Artificial Intelligence," Netflix, imdb.com/title/tt5960546/.

184 *And when you can feed it with tons of data:* Joel Hestness et al., "Deep Learning Scaling Is Predictable, Empirically," Baidu Research, December 2017, arxiv .org/pdf/1712.00409.pdf.

185 *With revelations that systems like Facebook:* Adam D. I. Kramer, Jamie E. Guillory, and Jeffrey T. Hancock, "Experimental Evidence of Massive-Scale Emotional Contagion Through Social Networks," *PNAS,* June 17, 2014, pnas .org/content/111/24/8788.

186 *we shouldn't be surprised:* "Blue Feed, Red Feed," *The Wall Street Journal,* graphics.wsj.com/blue-feed-red-feed.

187 *I first encountered Holmes's:* John Maeda, "Design in Tech Report 2017," SXSW, March 11, 2017, designintech.report/wp-content/uploads/2017/03/dit -2017-1-0-7-compressed.pdf.

187 *"Make a conscious effort":* Kat Holmes, *Mismatch: How Inclusion Shapes Design* (Cambridge, MA: MIT Press, 2018).

189 *more devices powered by Android:* Tripp Mickle, "Apple TV on a Samsung? iPhone Giant Makes Risky Jump to Other Devices to Sell Services," *The Wall Street Journal,* March 26, 2019, wsj.com/articles/apple-tv-on-a-samsung -iphone-giant-makes-risky-jump-to-other-devices-to-sell-services-115 53629788.

190 *The term "open source software":* Christine Peterson, "How I Coined the Term 'Open Source,'" Opensource.com, February 1, 2018, opensource.com /article/18/2/coining-term-open-source-software.

190 *difference between closed source and open source:* Olga Kozar, "Towards Better Group Work: Seeing the Difference Between Cooperation and Collaboration," *English Teaching Forum* 48, no. 2 (2010): 16–23, americanenglish .state.gov/files/ae/resource_files/48_2-etf-towards-better-group-work -seeing-the-difference-between-cooperation-and-collaboration.pdf.

192 *Microsoft surprised the world:* "Microsoft to Acquire Github for $7.5 Billion," Microsoft, June 4, 2018, news.microsoft.com/2018/06/04/microsoft-to -acquire-github-for-7-5-billion.

192 *and even Microsoft has announced:* Tom Warren, "Microsoft Is Building Its Own Chrome Browser to Replace Edge," *The Verge,* December 4, 2018, theverge .com/2018/12/4/18125238/microsoft-chrome-browser-windows-10-edge -chromium.

192 *Apple's web browser Safari:* Clint Ecker, "Apple Opens WebKit CVS Repository," *Ars Technica*, June 7, 2005, arstechnica.com/gadgets/2005/06/470.

193 *the latter is a GNU:* Footnote or reference that the reader might recall; GNU as "GNU's Not Unix" from the chapter one's discussion on recursion.

193 *"GPL license," which is more:* GNU General Public License, Free Software Foundation, gnu.org/licenses/gpl-3.0.en.html.

193 *computational ways to inspect these opaque AIs:* Sandra Wachter, Brent Mittelstadt, and Chris Russell, "Counterfactual Explanations Without Opening the Black Box: Automated Decisions and the GDPR," *Harvard Journal of Law & Technology* 31, no. 2 (2018), papers.ssrn.com/sol3/papers.cfm?abstract _id=3063289.

193 *more like "gray boxes":* Yury Makedonov, "Improved Testing of AI Systems with 'Grey-Box' Testing Technique," STARCANADA Conference, October 17, 2018, starcanada.techwell.com/program/concurrent-sessions/improve -testing-ai-systems-grey-box-testing-technique-starcanada-2018.

192 *efforts under way for AIs:* Kevin Hartnett and Quanta, "How a Pioneer of Machine Learning Became One of Its Sharpest Critics," *The Atlantic*, May 19, 2018, theatlantic.com/technology/archive/2018/05/machine-learning-is-stuck -on-asking-why/560675.

194 *I've compiled a list on:* This is a list of open source projects and for getting involved: howtospeakmachine.com/open.

Index